The Joneses

Of Nunawading Shire

Flower Growers to Generations of Melburnians

By

Roger K Jones and Elizabeth R Ogden

ISBN 978-0-9923-037-78

Printing 2019

Historic photos from family collections

Editor and publisher – Dana McCown

StoryBridge Press

Brisbane, Queensland, Australia

2019

rkjones9@tpg.com.au

liz.ogden@gmail.com

Yellow Roses on title page "Lady Hillingdon"

Front Cover design by Samuel Jesse Digital Creative

CONTENTS

Figure 1. James Jones (~1819 – 1899), father and grandfather of the Joneses of Nunawading Shire.

PREFACE

Liz Ogden and I, the authors, have often wondered about the origins of our love of plants and flowers. We both love plants – in fact, Liz continues to grow some of her father's orchid collection, and Roger had a 34-year career in CSIRO involved with plants in agricultural systems in the semiarid tropics of Australia and Africa.

We've discovered that an occupational involvement with plants and flowers can be traced back in the Jones family for at least three generations. Liz and I are both great great grandchildren of James Jones, a nurseryman in 19th century Buckinghamshire, England in the late 18th and early 19th century. So, for the purposes of this book, this nurseryman is our first generation of plant and flower growers. The nurseryman's son, also James Jones (~1818-1899) came to the Victorian goldfields in 1852 with his 12-year old son, James Frederick Jones. In 1862, James senior purchased a small farm in Preston on the outskirts of Melbourne, presumably with funds derived from his time on the goldfields. Marrying a second time in 1854, he listed his occupation as a pianoforte maker but he subsequently became a carpenter and builder. To date, we are yet to find evidence that he was involved with plants and flowers.

Four of James's six sons, however, were involved in some way or other in the plant and flower business. The eldest, James Frederick Jones (~1840-1926), the product of James's first marriage in England, worked for many years as a gardener for Theophilus Kitchen – a member of the wealthy candle and soap-making family of J Kitchen and Sons. Theophilus owned a large house with extensive gardens in the exclusive Melbourne suburb of Kew and James F. exhibited many plants and flowers from these gardens at the Richmond Horticultural Society where he was a prominent member. Another son, Stephen Alma Jones (1856-1938) rented land from his father for a time, presumably for plant production, then became a retail florist in Brighton, Victoria. A third son, Walter Edgar Jones (1860-1944) spent his whole life growing plants and flowers and this is where our story begins. The story then moves on to his youngest brother Arthur John Jones (1873-1965), our grandfather, who took over Walter's farm in 1917 and later purchased it. These four sons constitute our second generation of plant and flower growers.

In about 1935, Arthur retired, transferring and later selling the Jones flower farm to his two sons, Lionel Arthur Jones and Keith Erskine Jones, Liz and my father respectively; they constitute our third generation of plant and flower growers.

Roger K Jones

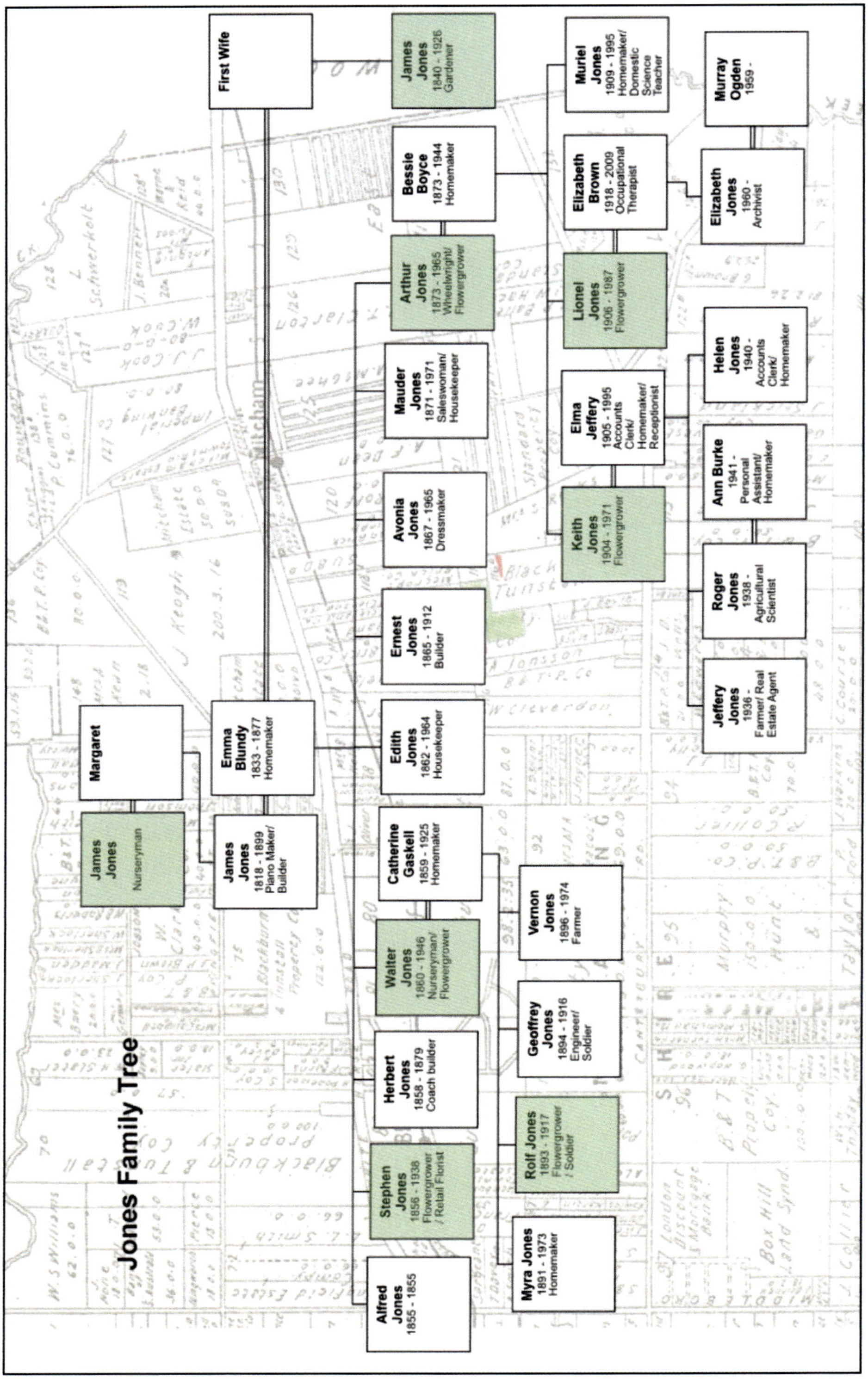
Jones Family Tree
First Wife
James Jones 1840 - 1926 Gardener
Margaret
James Jones Nurseryman
Emma Blundy 1833 - 1877 Homemaker
James Jones 1818 - 1899 Piano Maker/ Builder
Alfred Jones 1855 - 1855
Stephen Jones 1856 - 1938 Flowergrower / Retail Florist
Herbert Jones 1858 - 1879 Coach builder
Walter Jones 1860 - 1946 Nurseryman/ Flowergrower
Catherine Gaskell 1859 - 1925 Homemaker
Edith Jones 1862 - 1964 Housekeeper
Ernest Jones 1865 - 1912 Builder
Avonia Jones 1867 - 1965 Dressmaker
Mauder Jones 1871 - 1971 Saleswoman/ Housekeeper
Arthur Jones 1873 - 1965 Wheelwright/ Flowergrower
Bessie Boyce 1873 - 1944 Homemaker
Myra Jones 1891 - 1973 Homemaker
Rolf Jones 1893 - 1917 Flowergrower / Soldier
Geoffrey Jones 1894 - 1916 Engineer/ Soldier
Vernon Jones 1896 - 1974 Farmer
Keith Jones 1904 - 1971 Flowergrower
Elma Jeffery 1905 - 1995 Accounts Clerk/ Homemaker/ Receptionist
Lionel Jones 1906 - 1987 Flowergrower
Elizabeth Brown 1918 - 2009 Occupational Therapist
Muriel Jones 1909 - 1995 Homemaker/ Domestic Science Teacher
Jeffery Jones 1936 - Farmer/ Real Estate Agent
Roger Jones 1938 - Agricultural Scientist
Ann Burke 1941 - Personal Assistant/ Homemaker
Helen Jones 1940 - Accounts Clerk/ Homemaker
Elizabeth Jones 1960 - Archivist
Murray Ogden 1959 -

INTRODUCTION

This is a story of four families of the Joneses that supplied cut flowers, bulbs, and fruit tree seedlings, principally to the Melbourne wholesale and retail markets, over a period of about 70 years from 1890.

Such an enterprise required suitable land with easy access to markets, so the story begins with the early settlement of the eastern outskirts of Melbourne and the gradual subdivision of the original Crown Allotments into smaller and smaller parcels. It then goes on to describe the various land holdings of the three families, to provide some details of those families, and then to elaborate on the infrastructure, plant resources, and mode of operation of the businesses and their evolution over time, before making some concluding remarks.

Selection and Settlement of Melbourne's Eastern Outskirts

Crown land was first released for sale in the Parish of Nunawading in 1854 but it was not until after the passing of an Act commonly known as the Grant Act in 1865 that the area was settled. Through this Act, James Grant, the Minister for Lands aimed to shift land from the grip of squatters into the hands of small-scale settlers. Clause 42 of that Act impacted on settlement (in the Parish of Nunawading...authors) in an unexpected way.

Designed to help struggling miners, Clause 42 allowed for annual occupation licences to be issued for lots of up to 20 acres within ten miles of a goldfield. The "legal umbrella for land development in Nunawading" came by way of its proximity to the Anderson's Creek diggings in Warrandyte. As a result, "an explosion of settlement took place" in Nunawading but not by miners.

The 42nd settlers as they became known were mostly farmers but there were also wood carters, landless labourers, a gentleman, nurseryman, and a few described as illiterate – a farmer, labourer and farmer. Most of them came from the nearby districts of Box Hill, Hawthorn and Doncaster. They engaged in mixed farming, their lots "usually including dairies, fowl houses, piggeries, stables, barns and almost without exception one or more waterholes on each allotment". Their ventures were remarkably successful considering the soil they worked was regarded as among the poorest in Victoria.

An extract from Heritage Precincts Report 2004, City of Whitehorse, by S Westbrooke and J Dalrymple.

In 1864, the map of the Parish of Nunawading, County of Bourke shows that more than 80% of the ~10,000 acres of land east of the future Middleborough Road, north of the future Burwood Road and bounded to the East and North by Dandenong, Deep, and Koonung Creeks had been taken up, with the rest being either Crown land, Reserve, or Water Reserve (see Figure 2 and 3). At this point, there were only 31 landholders in this location, and allotments varied in size between 42 and 240 acres. The major landholders, each with more than 5% of this land, were: W Morton, P Riley, N Polak, J Holland, and W (William) Jones. (Allotment 116 that contained a future Jones flower farm was still crown land - see Figure 3.)

Twenty-eight years later, on the 1892 map of the Parish (see Figure 4), virtually all the names of these 31 landholders of 1864 had disappeared and were replaced by the names of more than 300 individuals, land syndicates, and companies. Allotments of 80 acres were common as it was possible to lease, and later purchase, up to four 20-acre allotments under the Grant Act.

By 1892, the surge of land speculation experienced in the 1880s (see following section) was over and the economic depression of the 1890s was underway.

Figure 2. Copy of portion of the 1864 map of Parish of Nunawading, County of Bourke lithographed at the Office of Lands and Survey, Melbourne, May 13th 1864 and reproduced in "The History of Nunawading" by Niall Brennan. It shows the ownership of Allotments east of the future Middleborough Rd. Crown Allotment 116 (outlined in pink) contains the future Jones flower farm. Note that the landowner, W. Jones, (highlighted in orange) was not a member of our Jones family.

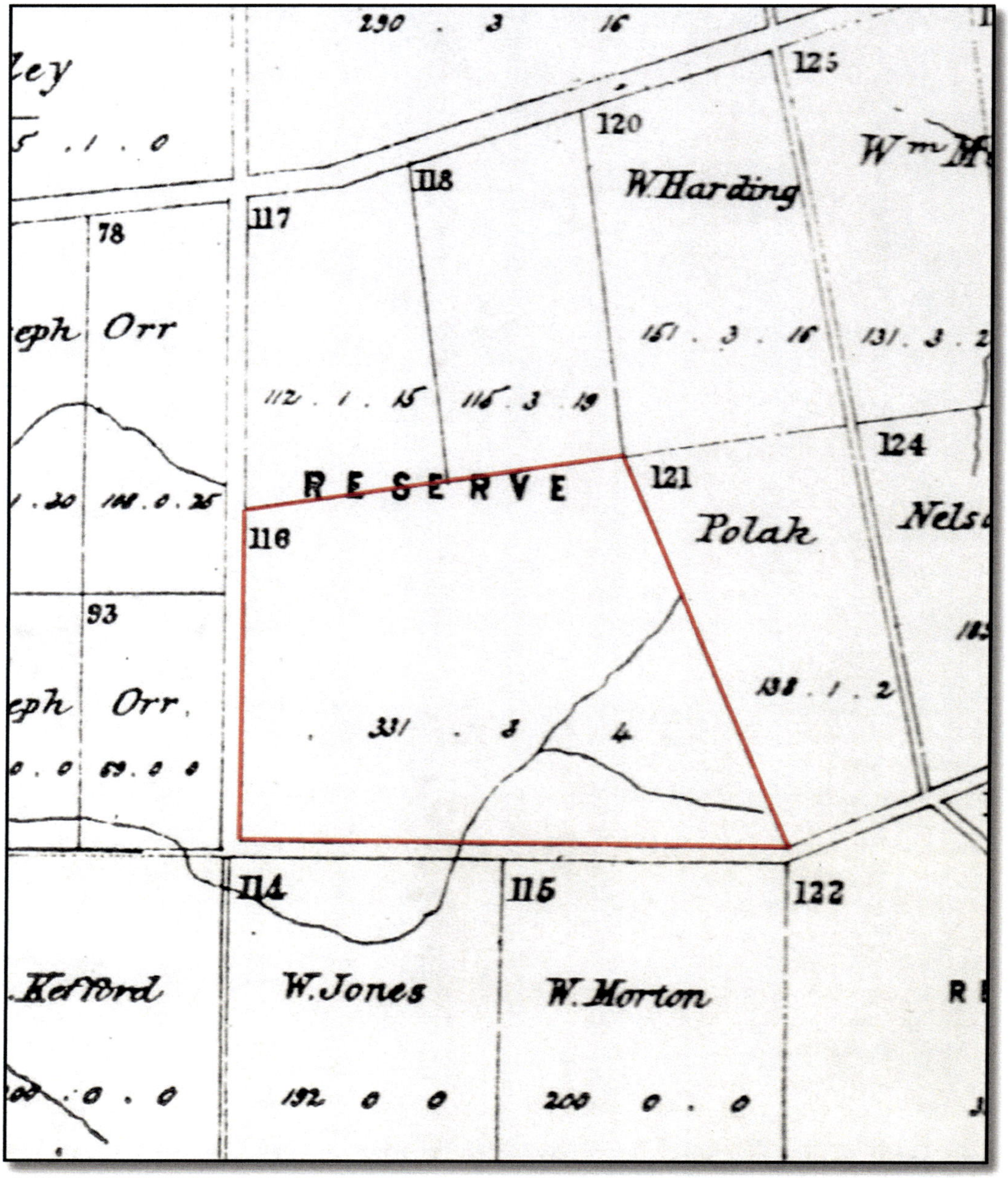

Figure 3. An even smaller portion of a Map of the Parish of Nunawading, County of Bourke published by the Office of Lands and Survey, May 13th 1864. The road to the south of the Allotment 116 was then referred to as Delaney's Rd and was to later become Canterbury Road; the road to its west was to later become Springvale Road. The road to the north of Allotments 117 and 118 was to later become firstly Whitehorse Road then its current name of Maroondah Highway.

Melbourne's 1880s Land Boom and 1890s Bust

Melbourne went through an incredible land boom during the 1880s that had an enormous influence on the development of the inner city itself and on the Shires at the outer margins of closer settlement, such as the Shire of Nunawading. So, it will be useful for the reader to understand the forces at play during this boom, and their impacts – both during and subsequent to it. The following extract introduces this topic:

> *From 1883 to 1889 Melbourne witnessed an extraordinary boom in real estate prices and land speculation. By 1889, the value of land in parts of central Melbourne was as high as that in London. In central Melbourne, huge sums of money were poured into opulent new office buildings, many for the building societies, land banks and mortgage companies that were driving the boom. These gothic buildings came to characterise central Melbourne, especially Collins Street, until the 1960s.*
>
> *The same speculators and financiers drove the expansion of the city at the edge, as market farms and orchards were swallowed up for new housing developments. The new suburban estates spread along the expanding rail and tramlines; prospective buyers of land were lured with the offer of free railway passes and lunches of chicken and champagne prior to the auction.*
>
> *The swathe of what are now middle suburbs were developed and consolidated during the 1880s, including Hawthorn, Camberwell and Box Hill in the east, Northcote and Brunswick in the north, Essendon and Footscray in the west and Brighton and Mentone in the south. Developers and estate agents promoted a suburban lifestyle that was embraced by both middle class and working class purchasers.*
>
> *Many of the land boomers, as the entrepreneurs were known, were gold-rush immigrants who had started as labourers, started small businesses, then joined temperance leagues, savings banks and other working-class self-improvement societies in the 1880s. The most successful entrepreneurs developed an intricate web of land banks, mortgage companies, and building societies, many with complex cross-ownership and Byzantine financial arrangements. As long as confidence held, the boom kept growing fuelled by speculation in land and shares in the companies.*
>
> *But once growth slowed and confidence stalled, the boom collapsed into an economic depression that was the most severe faced by Melbourne at that time. Melbourne would take decades to recover.*
>
> *The way in which the land boomers escaped their debts became as notorious as the dubious ways in which they had built their empires. Creditors were paid a small fraction of the sums they had invested, as little as a halfpenny in the pound (about one-fifth of a cent in the dollar). Many of these arrangements were made secretly and approved in the courts, enabling the boomer to escape the ignominy of bankruptcy and continue to operate other businesses that had been protected in the arrangements.*

Source: Gillespie, R. (2008) Land Boom in 1880s Melbourne in Museums Victoria Collections https://collections.museumvictoria.com.au/articles/2676.

Figure 4. Portion of a map of the Parish of Nunawading, County of Bourke in the Municipal Directory of 1892 (from Brennan) showing the ownership of Allotments east of Middleborough Rd. Relevant allotment Nos are underlined in red and land owned by WE Jones is coloured green. Allotment 113 on Springvale Rd contained WE Jones's first land purchase, whereas Allotments 116 D and 116 G, then both owned by the Blackburn and Tunstall Property Co, contained the principal future Jones flower farm.

LAND ACQUISITIONS BY THE JONES FAMILY

Land Acquisitions by Walter Jones

PRESTON

Walter Edgar Jones was born on April 18, 1860, the son of James Jones and Emma Jones, formerly Blundy. Walter's father, James, had purchased 5½ acres of land on or about Raglan and Newcastle Streets, South Preston, Victoria in 1862. This is where Walter and his siblings were raised. We don't know where Walter and his older brother Stephen gained their early nurseryman experience but we do know (from a thumbnail sketch of James Jones, probably from information James himself supplied, in an 1888 Melbourne promotional publication[1]) that at least one of his sons was in the plant and flower business.

On his return to Victoria from New Zealand, where he was engaged in mining and contracting for two years, he (James) purchased 5 1/2 acres of land in South Preston in 1862. He has also 4 1/4 acres at North Preston, laid out as an orchard and flower-garden and let to his son.

The Rate Books from that time throw more light on the land ownership and lease arrangements. In 1886/87, Stephen and Walter, both described as gardeners, were paying rates on a portion of the Raglan St land owned by their father, James Jones. By 1890, Stephen, now described as a gardener and florist, was paying rates on two blocks – one that included a house, owned by his father, and another block owned by Moses Blundy. (In 1892, the Sands and McDougall's Post Office Directory for Melbourne shows James Jones living at 70 Hodgkinson St Clifton Hill, so he may well have moved out of the original family home and leased it to his eldest son from about 1890.)

Walter, described then as a jeweller, was also paying the rates on two blocks – one owned by his father and another that he owned that included a house. (Walter may have worked for a time for his wife's brother who had a jeweller shop or workshop at 10 Royal Lane, off 210 Little Collins-St, Melbourne.) The latter block is probably the small suburban block of land (0.11 acres) in Raglan Street, South Preston that Walter purchased from his uncle, Moses John Blundy, on 21 December 1889. It was probably here that Walter and his new wife, Catherine, née Gaskell, commenced married life in 1890.

---

1 *"Victoria and its Metropolis - past and present (1888). Volume IIB: The Colony and its People in 1888". Published by McCarron, Bird & Co 1888. From page 735, chapter on "The outlying suburbs of Melbourne", Section on Builders, Timber Merchants etc. State Library of Victoria.*

TALLY-HO/ EAST BURWOOD

It seems that Walter, aged about 32 years, purchased his first 20 acres of land for a nursery at Tally-Ho or East Burwood on 12th of April 1892. It was within Allotment 113 on the eastern side of Springvale Road (formerly Hunt's Lane) just north of the future Burwood Road, now Burwood Highway (see Figures 4, 5, and 6). Walter may have received financial assistance from his father to purchase this Springvale Rd land, as is suggested by the fact that both Walter and his brother Stephen received the sum of only one shilling in their father's 1899 Will. However, additional funds would have been required to build a house for his family and establish a nursery, so this is presumably the reason that he mortgaged (and subsequently redeemed) this land in April 1892 and again in April 1898. Three of Walter and Catherine's 4 children were born in Preston with the third, Geoffrey Ernest, being born in 1894. Vernon, the youngest, was born in Burwood in 1897 probably not long after they shifted there.

Tally-Ho was close to the boundaries of the Nunawading and Mulgrave Shires – the dividing line being Highbury Road. A store had been built at the corner of Highbury and Blackburn Roads in the early 1860s, a primary school had opened nearby in 1861 for children from families in each Shire, and a Hotel had been built next to the store in 1871. Over the next thirty years, farming settlements, particularly orchards, had developed along Highbury and Springvale Roads and a local Fruit Growers Association had been formed. In about 1902 the Methodist Church had established the Tally-Ho Boys Home. *(Information from http://www.victorianplaces.com.au/tally-ho.)*

The locality was also described in the Australian Handbook of 1903 as follows:

> **TALLY-HO** (37° 45′ S. lat., 140° E. long.) is a postal centre in the county of Bourke, electoral district of Oakleigh, police district of Burwood, shire of Oakleigh, 11 miles E. of Melbourne, on boundary road dividing Oakleigh and Nunawading. Communication with capital: horse to Burwood, cab to Camberwell, rail to Melbourne. Hotels: Tally-Ho and Boundary. State school (No. 454). Church: Primitive Methodist. Good orchard land, with fine clay subsoil, and splendid brick and tile clay. Formation: schistose. Population, about 450.

Unfortunately, we do not know exactly what sort of plant-related business Walter ran here over about the next ten years. One map and the land title showed him as a florist but a Ledger from the time held by one of the authors records the dispatch of extensive consignments of young fruit trees, presumably as bare rooted or balled plants, to landholders in the Natal region of South Africa in about 1902-03 suggest he was actually a nurseryman (see Figure 7). There were over 400 consignments consisting of named varieties of apples, pears, apricots, peaches, nectarines, plums, persimmons, oranges, mandarins, cherries, pomegranates, olives, currents, gooseberries, hazel nuts, quinces, and figs. He may also have supplied such fruit trees to the burgeoning apple and pear orchards in the vicinity of Tally-Ho.

Entered in the Register Book
CANCELLED Vol 1604 Fol 320696

VICTORIA.

Certificate of Title,

UNDER THE "TRANSFER OF LAND STATUTE."

George Nicholson Taylor of Collins Street West Melbourne Bank Manager
is now the proprietor of an Estate in Fee-simple, subject to the Encumbrances notified hereunder in All that piece of Land, delineated and coloured red on the Map in the margin, containing twenty [illegible] acres or thereabouts being part of Crown portion one hundred and thirteen parish of Nunawading County of Bourke Together with a right of carriage way over the road delineated and colored brown on the said map

Dated the second day of September One thousand eight hundred and eighty four

CANCELLED

Assistant Registrar of Titles.

ENCUMBRANCES REFERRED TO.

ORIGINAL CERTIFICATE
Not to be dealt with outside the Titles Office

T01604-696-1-9

Figure 5. An excerpt from the Certificate of Title (Vol 1604, Fol 320696) for the two blocks of land on Springvale Road, East Burwood owned in 1884 by George Taylor. In 1891 (see Figure 6), he transferred it to the Land Credit Bank of Australasia Limited, and they in turn transferred it on April 12, 1892 to Walter Jones. (The road on the eastern boundary over which the property had carriage rights is probably the present-day Stanley Rd., Vermont South.)

Information courtesy of Victoria's Dept of Environment, Land, Water & Planning (DELWP) via the Public Records Office (PROV).

We presume that access to the railway line for easier dispatch of his produce to market led Walter, early in the twentieth century, to sell his land in Springvale Road, East Burwood and look for land closer to Mitcham and Tunstall Railway Stations. This line to Ringwood had opened in about 1888. Hence, Walter sold his portion of Allotment 113 on November 12, 1903.

(In 2018, the 20 acres of land that Walter purchased is now covered with housing that faces a number of streets in the suburb of Vermont South, including Hawthorn Rd, Cascade Drive, Elonara Rd and Stanley Rd.)

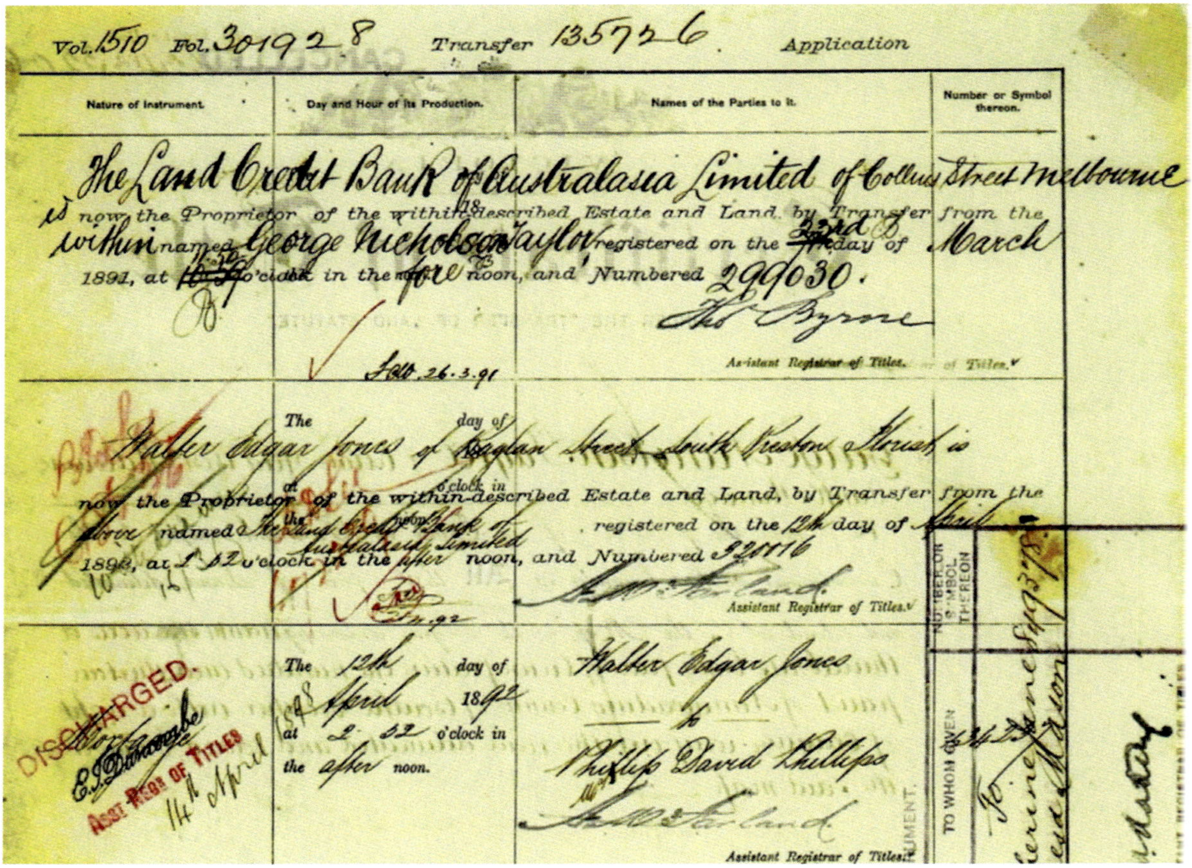

Vol. 1510 Fol. 301928 Transfer 135726 Application

Nature of Instrument	Day and Hour of its Production	Names of the Parties to it	Number or Symbol thereon

The Land Credit Bank of Australasia Limited of Collins Street Melbourne is now the Proprietor of the within described Estate and Land by Transfer from the within named George Nicholson Taylor registered on the 23rd day of March 1891, at o'clock in the fore noon, and Numbered 299030.
Thos Byrne
Assistant Registrar of Titles.

The day of
Walter Edgar Jones of Raglan Street South Preston Street is now the Proprietor of the within described Estate and Land, by Transfer from the above named The Land Credit Bank of Australasia Limited registered on the 12th day of April 1892, at 2.52 o'clock in the after noon, and Numbered 320116
McFarland
Assistant Registrar of Titles.

The 12th day of April 1892 at 2.52 o'clock in the after noon.
Walter Edgar Jones to Phillip David Phillips
McFarland
Assistant Registrar of Titles.

DISCHARGED
ASST REGR OF TITLES
14th April 1898

Figure 6. A further excerpt from the Certificate of Title (Vol 1604, Fol 320696) for 20 acres of land on Springvale Road, East Burwood showing that Walter Jones purchased it 12 April 1892, immediately mortgaged it to Phillip David Phillips and subsequently discharged that mortgage on 14 April 1898.

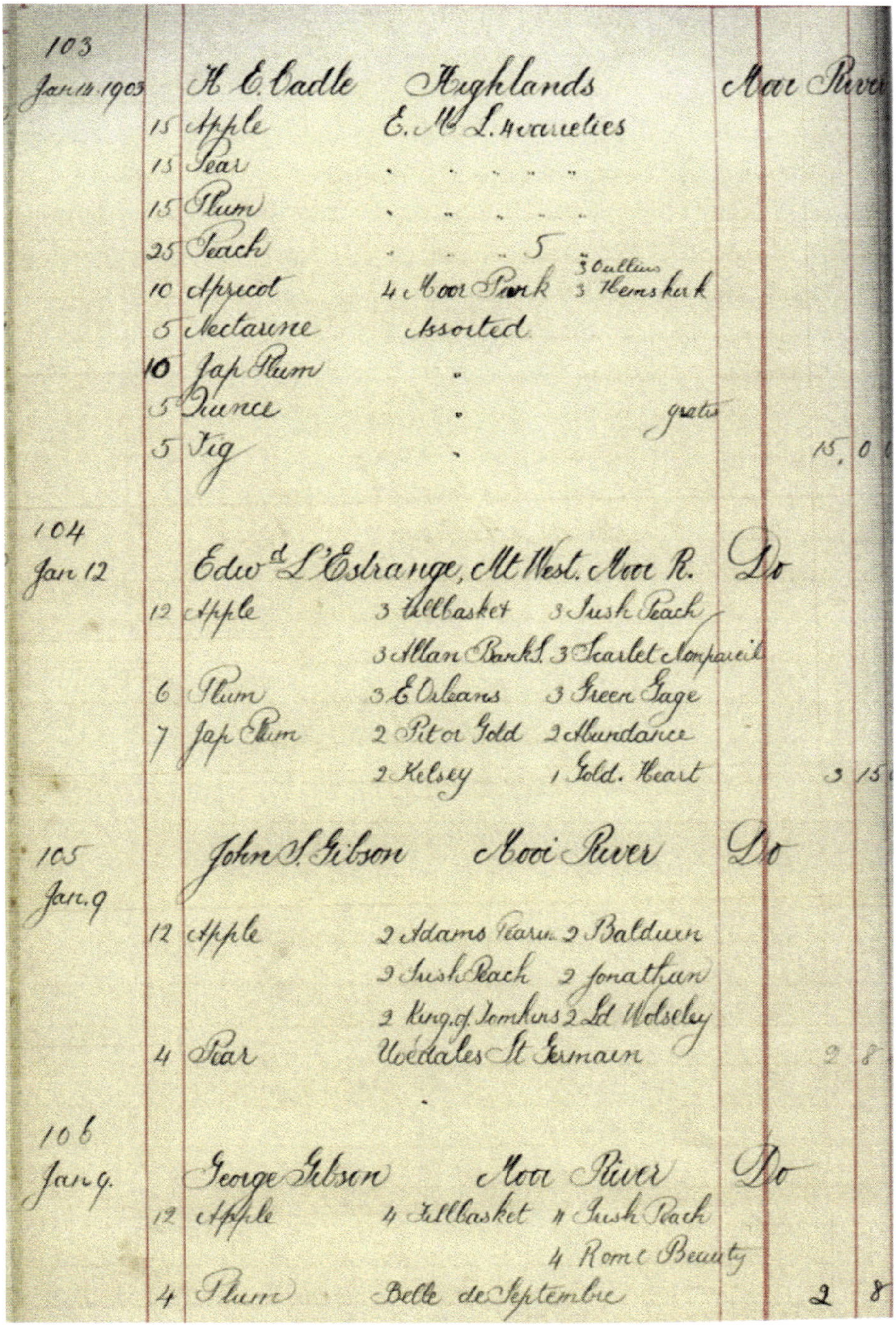
103
Jan 14 1903 H. E. Cadle Highlands Mooi River
15 Apple E. M. L. 4 varieties
15 Pear " " " " "
15 Plum " " " " "
25 Peach " " " 5
10 Apricot 4 Moor Park 3 Oullins 3 Hemskirk
5 Nectarine Assorted
10 Jap Plum "
5 Quince " gratis
5 Fig " 15. 0

104
Jan 12 Edwd L'Estrange, Mt West. Mooi R. Do
12 Apple 3 Fillbasket 3 Irish Peach
3 Allan Bank 3 Scarlet Nonpareil
6 Plum 3 E Orleans 3 Green Gage
7 Jap Plum 2 Pit or Gold 2 Abundance
2 Kelsey 1 Gold. Heart 3 15

105
Jan 9 John S. Gibson Mooi River Do
12 Apple 2 Adams Pearm. 2 Baldwin
2 Irish Peach 2 Jonathan
2 King of Tomkins 2 Ld Wolseley
4 Pear Uvedales St Germain 2 8

106
Jan 9. George Gibson Mooi River Do
12 Apple 4 Fillbasket 4 Irish Peach
4 Rome Beauty
4 Plum Belle de Septembre 2 8

Figure 7. Excerpt from a page of the Dispatch Ledger of W.E. Jones showing consignments of fruit tree seedlings that he had dispatched to landholders in the Natal District of South Africa in 1903.

TUNSTALL

It appears that Walter Jones became interested in moving his operations closer to Tunstall and Mitcham early in the twentieth century. While his first land purchase in Mt Pleasant Rd (formerly known as Boyle's Rd) was not officially registered to him on the title until 1904, it appears that he was the ratepayer on the property from at least October 1901. This is supported by both the rate records and an account in the Reporter (Box Hill) of Friday 7 March, 1902:

> ***NUNAWADING SHIRE COUNCIL.*** *At the ordinary meeting of the above council on Tuesday evening there were present: Councillors Hunter (president), Andrew, Smith, Horkings, Garrett, Walker, Cox, Hall, Morton, Fankhauser, Bayley and Aspinall.*
>
> *Correspondence: From C. Course, W. Boyle, H. Furborough, J. Boyle, A. Williamson, A. H. Plumridge, and W. E. Jones, ratepayers on Mount Pleasant road, drawing attention to the bad state of the entrance from that road to Canterbury road. On the motion of Crs Hall and Walker, it was agreed that £5 be expended in putting in the necessary crossing.*

Interestingly one of the authors (Roger) remembers this entrance from Mt Pleasant Rd. to Canterbury Rd. as still being in a poor state 50 years later – in the early 1950s!

On 14 April 1904, Walter Edgar Jones formally purchased a 6.66-acre block of land from the Freehold Investment and Banking Company of Australia Ltd (in liquidation). It fronted an easement that was to become Mt Pleasant Road and was a subdivision of the original Allotment 116 of 331 acres, and subsequently of Allotment 116 G of about 61 acres. Purchasers of other neighbouring blocks between 1900 and 1906 were: Charles Course, Alfred John Williamson, Arthur Henry Plumridge, Joseph Boyle, and Ambrose Fredrick Thiele. These were to become the long-term neighbours of the farm and well known to the Joneses.

On the same day and from the same company, Walter purchased another block of 7.71 acres that was originally part of the neighbouring Allotment 116 D. This block had no easements for access to the developing road network. However, as it shared a short boundary of one chain (20.1 m) with the other block, the two were amalgamated into one of 14.37 acres as shown in Figure 9.

We believe that the western-most block that faced Mt Pleasant Rd was only partially cleared and had no infrastructure when purchased by Walter. We knew it as the "bottom farm or garden". The eastern-most block on the top of the hill was part of the lane originally owned by Peter Boldini (see Figure 8). We knew it as the "top farm."

SATURDAY, 24th FEBRUARY.

At Half-past 4 0'Clock.

On the Property.

MOUNT PLEASANT,
12 Miles from Melbourne, and 1 3/4 Miles from Mitcham
Railway Station,
On the Lilydale Line,
CANTERBURY-ROAD,
Close to Church, Post Office and State School,
80 ACRES LAND, On the Crown of the Hill, View of the Bay.
WEATHERBOARD COTTAGE,

Four Rooms, Cemented Tank, Outhouses, Stabling.
Title Freehold.
BEAUCHAMP BROTHERS have received instruct-
tions from Mr. Peter Boldini to sell by auction,
on the property, on Saturday, 24th, at half-past 4 o'clock,
His very comfortable
FARM, situated one mile from the Mitcham station,
by the paddocks, or 1 3/4 miles by the roads.
The position is charming, commanding a
grand view of the bay, being on the crown of
a hill. The locality is well known, and recom-
mended for its salubrity.
80 ACRES LAND,
all well fenced with posts and rails, live
hedge in front, three paddocks fenced with
acacia hedge, fairly timbered, 25 acres under
cultivation, three acres orchard, containing
some of the best fruit trees in the colony.
SUBSTANTIAL WEATHERBOARD COTTAGE,
containing four good-sized rooms, also out-
houses, stabling, large cemented tank.
Good fishing and shooting close to the property.
A never-failing creek runs through portion of the land.
Stock and working plant can be taken at a valuation.
Train leaves Melbourne at 20 minutes to 4 o'clock,
and arrives in time for the sale.
Plan and fuller particulars of the property obtain-
able from the auctioneers.
Title, Crown grant.

Figure 8. Transcript of an advertisement in the Melbourne Age for Peter Boldini's 80-acre property at Mt Pleasant. (Source: The Age (Melbourne, Vic 1854-1954), Saturday 10 February 1883, Page 2.

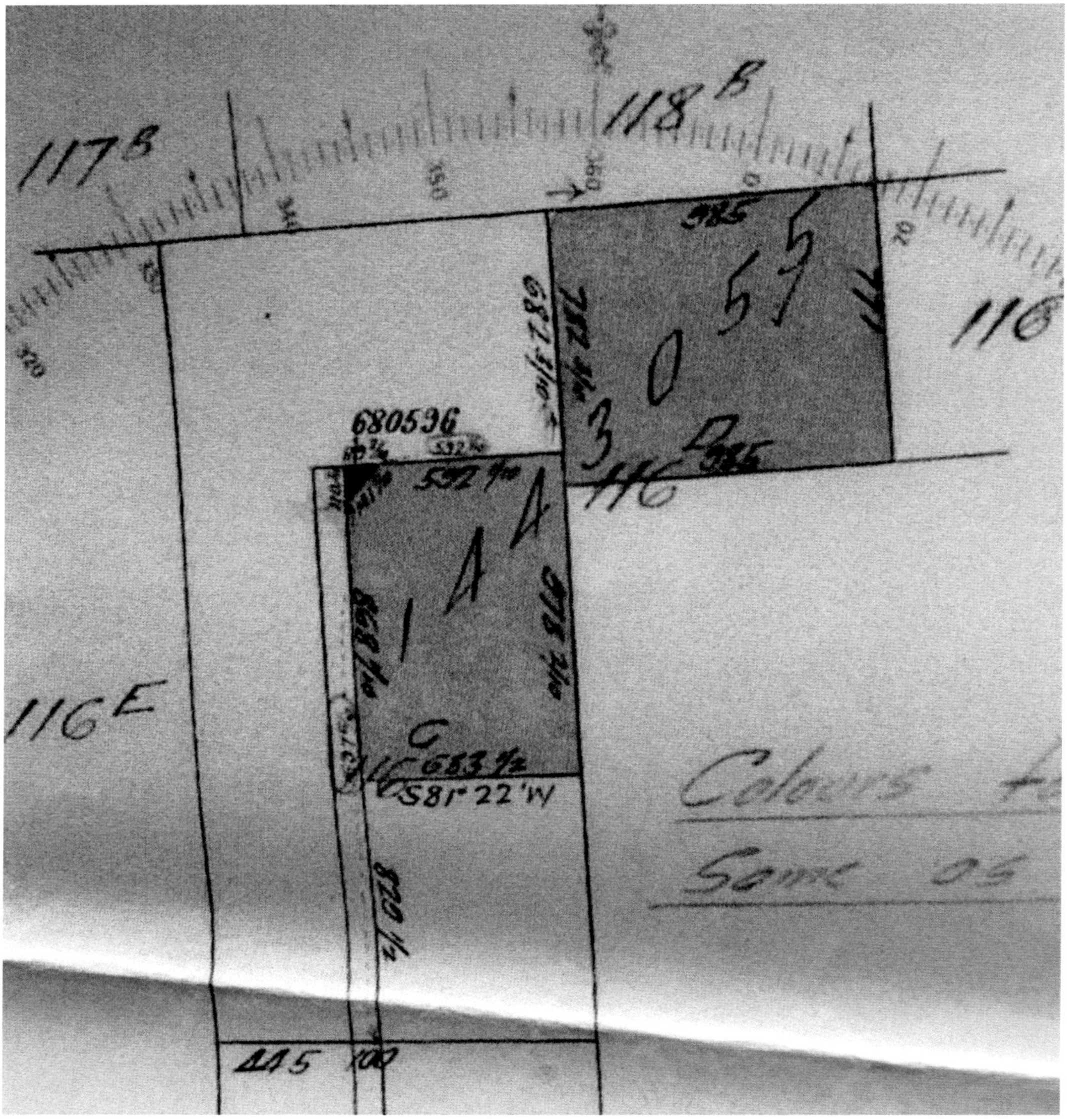

Figure 9. An excerpt from the 1904 title documents (Vol 2983, Fol 596471) for the two blocks of land purchased by Walter Jones and amalgamated to become the original "Rosemont" flower farm. The most westerly block fronts the future Mt Pleasant Rd that was developing from the south. (Measurements on sides of blocks are in links.)

Information courtesy of Victoria's Dept of Environment, Land, Water & Planning (DELWP) via the Public Records Office (PROV).

The Top Garden

The earlier ownership and subdivision history of the "top garden" is of particular interest because it became the principal hub of the farm and may have provided some of the infrastructure, so we present a summary here.

The "top garden" was originally part of Allotment 116 D that Patrick Geraghty purchased from the Crown in May 1871 for £81. This approximately 83-acre allotment was bounded on the south by the future Canterbury Road. Patrick mortgaged it several times in the early 1870s to the Land Mortgage Bank of Victoria Ltd, then transferred it to William Geraghty, presumably a family member, in May 1877.

Then, in August 1877, Peter Boldini, a vigneron, purchased the Allotment. A search of the Municipal Rate Books for the time confirms that Boldini, shown there simply as "a farmer", was the owner from at least 1879 to 1884. Boldini, however, soon seems to have got into financial trouble that required him, between 1877 and 1882, to mortgage the land several times to various individuals and then to the Australian Deposit and Mortgage Bank Ltd. Finally, in February 1883, he advertised the farm in the Melbourne Age for sale by Auction (Figure 8). The auction was apparently unsuccessful but the land was finally sold or transferred in August 1885 to the Freehold Investment and Banking Company Australia Limited.

Clearly, these were very turbulent times!

This Freehold Investment and Banking Company Australia Limited held the land until 14 April 1904 when, in receivership, it sold a small northern section to WE Jones. The two portions of land were then amalgamated into one of 14.37 acres (Figure 9). As noted above.

[Appendix 1 gives full details of the ownership and progressive subdivision of Allotments 116 G (containing the "bottom garden") by David Boyle and his family. It also contains a more complete description than presented here of the ownership and progressive subdivision of Allotment 116 D (containing the "top garden")]

MISCELLANEOUS OTHER LOCATIONS

On July 6 1911, Walter received his Certificate of Title for two blocks of land totalling 6 acres on the eastern side of Springvale Rd, Tunstall. These were located on land in Allotment 117A formerly owned by P. Walker. They commenced about 170 metres north of the northern boundary of Allotment 116 B in the vicinity of the present-day East India Avenue. Two years later, however, on May 30, 1913, he sold the southernmost block of about 3 acres to John Malcolm Roberts but retained the other until 1921 (see below).

Interestingly, four years later, on June 30, 1915, Walter received his Certificate of Title for another four blocks of land totalling about 4.3 acres in Mt Pleasant Rd, Tunstall. Two of these, totalling about 2.1 acres, on the west side of Mt Pleasant Rd, in the vicinity of the present-day East India Avenue, were almost back-to-back with the above-mentioned block fronting Springvale Rd that he still owned. Was this by chance or design? Unfortunately, we do not know whether these two blocks, totalling over 5 acres, were regarded by Walter as potential flower growing land, nor do we know whether Walter or his sons ever farmed them.

Walter sold these two blocks in 1920 and 1921 when retiring and disposing of his Tunstall land holdings. The remaining 2.25 acres were on the northern side of the short east-west section of Mt Pleasant Rd. These were sold to his brother Arthur.

Walter also purchased a number of other potential residential or commercial blocks of land in the Tunstall and Mitcham districts in the 1912 to 1914 period, presumably as investments. Thus: On 29 May 1912, he purchased two blocks in the heart of Mitcham on the corner of Whitehorse Rd and Station St. He sold these in 1923; one became the site of the Mitcham Post Office, and the other of the Mitcham Branch of the State Savings Bank of Victoria.

On 29 May 1914, he purchased 0.4 acres of land at the corner of Whitehorse Rd and Walkers Rd, in the vicinity of O'Shannessy St, from The Argus Permanent Building and Investment Society. With an eye possibly on his later retirement, Walter also purchased a number of adjacent blocks of land in Healesville between 1909 and 1911.

Land Acquisitions by Arthur Jones

Arthur John Jones, Walter's youngest brother, born in 1873, took over the management of the Flower Farm of about 14.4 acres in 1917. He presumably paid it off over the next decade, becoming the official Title Holder on 31 December 1929. In addition, Arthur purchased the remaining 2.25 acres of land in Mt Pleasant Rd referred to above on November 16 1922. This land had not been farmed and remained well wooded until the late 1940s or early 1950s.

The 1926/27 period was clearly one of major subdivision of Tunstall/ Mitcham land. Land prices were relatively cheap, so Arthur and/ or his wife invested in a number of small residential blocks that became available. Thus:

- Arthur purchased a block of land at the corner of Rooks Rd and New St, Mitcham on 9 August 1919. This is located close to the bend in Rooks Rd near its junction with the present-day Lorikeet St. He sold it a year later.
- Arthur also purchased three blocks of land on 6 January 1925, part of the Premier Estate of Mitcham that was first advertised in December 1919. The three contiguous blocks were on the corner of Halls Parade and Premier Ave. They were sold between 1948 and 1951.
- Arthur's wife, Bessie Edith Jones, purchased a small block on the western side of Shady Grove, Tunstall on 2 March 21, 1927. [Some of the other purchasers of the Shady Grove blocks around that time were from the Joyner, Boyle, and Chugg families that were well-known to the Joneses.]

In the 1930's Arthur and Bessie's sons, Keith and Lionel gradually took over the further development and running of the business. Arthur also encouraged his daughter Muriel to invest in property so, in September 1934, she purchased a house at 91 Bluff Rd, Black Rock, a seaside suburb of Melbourne. Soon after, Arthur and Bessie retired and went to live in this house. Bessie died in 1944.

Arthur subsequently married Beatrice May Hunter, a widow of Springvale Rd, Tunstall, in 1945 and had a new house built on portion (now 65 Mt Pleasant Road) of the 2.25-acre block of land that he had purchased from his brother in 1922 (see above). Over the next ten years he sold off the remainder of the 2.25 acres as residential blocks and named the street so created, Erskine St.

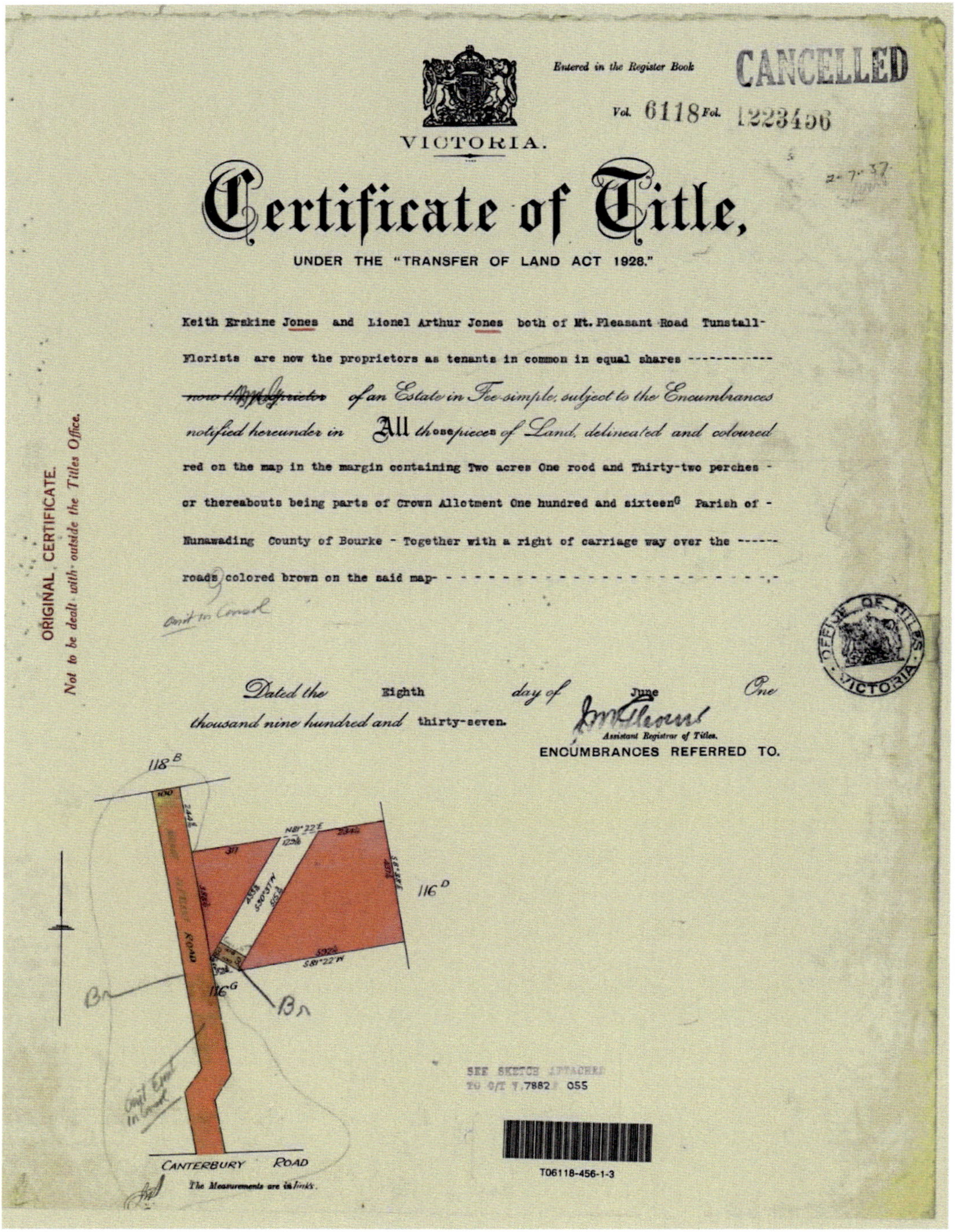

Entered in the Register Book

CANCELLED

Vol. 6118 Fol. 1223456

VICTORIA.

2-7-37

Certificate of Title,

UNDER THE "TRANSFER OF LAND ACT 1928."

Keith Erskine Jones and Lionel Arthur Jones both of Mt. Pleasant Road Tunstall- Florists are now the proprietors as tenants in common in equal shares ------------ ~~now the proprietor~~ of an Estate in Fee-simple, subject to the Encumbrances notified hereunder in All those pieces of Land, delineated and coloured red on the map in the margin containing Two acres One rood and Thirty-two perches - or thereabouts being parts of Crown Allotment One hundred and sixteen G Parish of - Nunawading County of Bourke - Together with a right of carriage way over the ------ roads colored brown on the said map- -.-

Dated the Eighth day of June One thousand nine hundred and thirty-seven.

Assistant Registrar of Titles.

ORIGINAL CERTIFICATE.
Not to be dealt with outside the Titles Office.

OFFICE OF TITLES VICTORIA

ENCUMBRANCES REFERRED TO.

SEE SKETCH ATTACHED TO C/T 7882 055

T06118-456-1-3

Figure 10. Excerpt from the Certificate of Title (Vol 6118 Fol 1223456) for a parcel of 2.45 acres of land from within the estate of Oswald John Boyle. Keith and Lionel purchased it on 8 June 1937.

Information courtesy of Victoria's Dept of Environment, Land, Water & Planning (DELWP) via the Public Records Office (PROV).

Land Acquisitions by Keith and Lionel Jones

In the 1920s, both Keith and Lionel followed their father's lead and purchased small blocks of land in the Tunstall district, presumably as investments when prices were favourable. Details are as follows:

- Keith Erskine Jones purchased a block in South St (now Wood St) Tunstall near the junction with West St. on 15 May 1925. This is just to the south of the present Nunawading Shopping Centre. Keith sold this land in 1947.
- Lionel Arthur Jones purchased a block on the southern side of Whitehorse Rd just to the east of its intersection with Springvale Rd Tunstall on November 26, 1926. He sold the land in 1953.

Arthur obviously intended his sons, Keith and Lionel, to take over the "Rosemont" flower business in due course – in fact they were running the business from about 1935. Arthur finally sold "Rosemont" to his two sons, operating as an informal business partnership, at the end of the Second World War. The contract, dated 1 September 1945 was for £2,000, with payment spread over seven years at £260 per annum and with one final payment of £80 at the expiration of the eighth year. There was a Caveat on the land from 26 October 1945 to 8 June 1953 when it lapsed. It was presumably taken out by Keith and Lionel to protect their interests as they gradually paid their father for the property. The title was transferred formally to the two brothers on 8 June 1953.

In 1937, an opportunity arose for Keith and Lionel to jointly purchase a parcel of land adjacent to "Rosemont" (see Figure 10). It consisted of two acres, one rood and 32 perches (2.45 acres), part of Crown Allotment 116 G that was originally owned by David Boyle and was subsequently transferred to Oswald John Boyle. The land was bisected by a narrow strip of land, 1-chain, (22 yards or 20.1 m) wide, owned by the Melbourne Metropolitan Board of Works (MMBW). Buried in this strip were 3-foot diameter pipelines bringing water from a Reservoir in Mitcham to a Water Tower on a hilltop in Mont Albert, and thence for distribution to suburbs of Melbourne. Keith and Lionel presumably purchased the land to provide scope for future expansion of the farm's operations, or as an investment.

By the mid-1950s, the inexorable spread of suburban Melbourne began to engulf Nunawading. The activities of developers and steep increases in Council rates put great pressure on existing farming and orcharding businesses in the District. In the case of the Jones Flower Farm, it was the State Government in the form of its Education Department that started the process. After a series of negotiations, the Department on 29th April 1957 issued a "Notice to Treat" to resume 5 acres 3 roods and 28 perches of land (5.93 acres) from the "bottom garden". This took about 95% of the "bottom garden", but it allowed Keith and Lionel Jones to retain a narrow access corridor to the "top garden". We have no record of the price that the Department paid for the acquisition. The Department subsequently erected a Primary School on the land that opened as School No 4808 in September 1958. It was known as the Nunawading South State School. The school was rebuilt in 2014/2015 and is now known as the Mount Pleasant Road Nunawading Primary School.

Acquisition of about half of the farm further reduced the viability of the flower growing business, so Keith and Lionel sold the remainder of the farm to a developer in about 1958 and vacated the farm in about 1960, having sold off as much infrastructure and as many plants as they could in the previous 12 months.

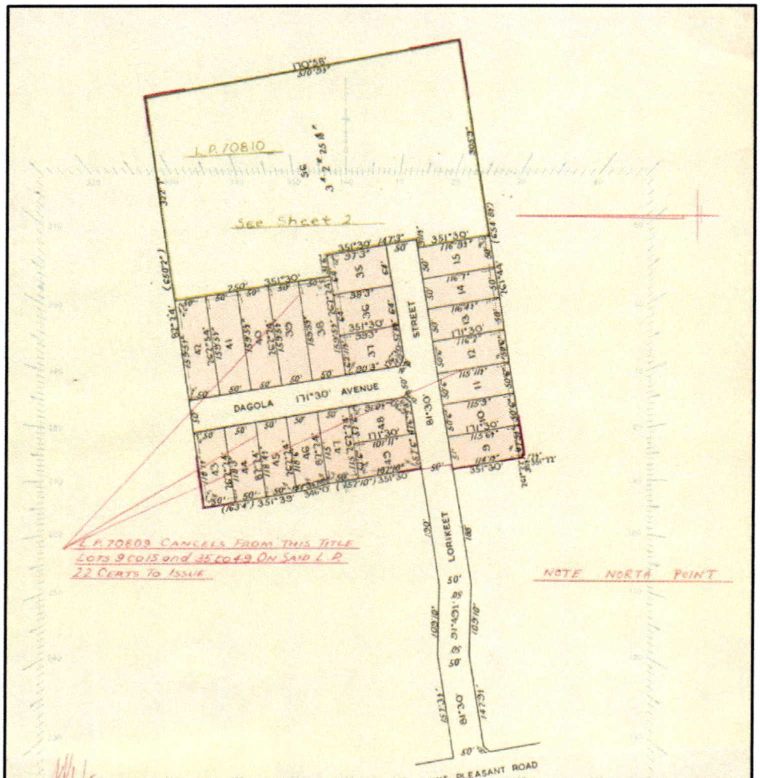

Figure 11. Excerpts from the Certificate of Title (Vol 8425, Fol 261) for subdivision of parts of the"top garden" of the Rosemont Flower Farm.

Information courtesy of Victoria's Dept of Environment, Land, Water & Planning (DELWP) via the Public Records Office (PROV).

This brings to a close the story of the various land holdings of the three Jones families over 70 years of flower growing in the Tunstall/ Nunawading District.

THE THREE JONES FAMILIES

Walter and Catherine

Figure 12. Believed to be Walter circa 1920.

After about 10 years in Tally Ho/ East Burwood, Walter Jones and his wife Catherine (née Gaskell), both aged about 44 years, moved to the Tunstall farm in the early 1900s with their four children – Myra Gaskell, Rolf Stanley, Geoffrey Ernest, and Vernon Edgar Jones aged approximately 11, 8, 7, and 6 years respectively.

We believe that Walter was the principal long-term nurseryman and flower grower rather than his brother, Arthur. In other words, we think that Walter was the "Builder", and that it was he who developed the farm's infrastructure and greatly expanded the flower growing and bulb production business. Among many other activities, Walter grew dozens of varieties of Daffodils or Narcissus and was a breeder or at least a selector of unusual types that he could exhibit at flower shows. He named a few himself, e.g. Nestalla (after his property at Healesville) and Anzac.

The children transferred to Mitcham State School and Walter was an active member of the School Committee and eventually chaired it from 1914 to 1917. Walter and Catherine were also both very active in other local community affairs. Walter stood unsuccessfully for election in the Mitcham Riding of the Nunawading Shire Council in 1908 and 1911. He therefore put his efforts into the Mitcham and Tunstall Progress Association, becoming President in 1914, having previously served as Vice President and committee member. (His 22-yr old son, Rolf, was the Secretary until he enlisted.) Walter was also on the Parish Council of the Mitcham's Church of England for one or more years.

Not to be outdone, Catherine was active in the formation of the Mitcham Patriotic Fund at war's outbreak in 1914, was its first President, and contributed heavily to its fund raising over the next four years. In mid-1916, however, tragic circumstances forced Walter and Catherine to review their lives. Two of their sons were killed in the same World War 1 battle in north-eastern France. (see box following page)

In 1917 therefore, Walter, then aged about 57 years, retired initially to Healesville, and sold the flower farm to his younger brother, Arthur.

Figure 13. Rolf Stanley Jones AWM HO6651

Figure 14. Geoffrey Ernest Jones AWM HO6658

On July 19, 1916, in the northern French village of Fromelles, Australia suffered its worst-ever military defeat when a British officer ordered 15,000 of our best and bravest to go 'over the top' and attack the German lines. Eight hours later, more than 5,500 Diggers lay dead or wounded: the equivalent of all Australian casualties from the Boer, Korean, and Vietnam wars combined.

Geoffrey Ernest Jones, b 14 August 1894, an engineer and the younger of the two Jones brothers, was the first to enlist. He embarked on 21 October 1914 and joined the Gallipoli campaign of the Australian Imperial Forces. He was wounded on 25 April, the first Anzac Day, was later evacuated to Egypt but rejoined his unit of the 60th Battalion at Gallipoli in late May, and remained there until early 1916. On 18 June 1916 he embarked at Alexandria to join the British Expeditionary Forces in France. By then a sergeant in charge of 7 Platoon, B Company, Geoffrey was killed at Fromelles in north-eastern France on 19 July 1916, aged 21 years. His remains were never found but may be in a mass grave at VC corner near the site of the battle.

Geoffrey's brother, Rolf Stanley Jones, b 17 November 1892, a Private in the 59th Battalion, embarked on 11 October 1915 – a year after his younger brother, stopped briefly in Egypt, then sailed for Europe on the same ship as his brother. Soon after arrival in France he was wounded at Fromelles (his nose was blown off) also in the same battle as his brother. Rolf was evacuated for treatment to England where he unfortunately died of meningitis at the Cambridge Hospital, Aldershot, Hampshire on 15 February 1917, aged 23 years. He was buried initially at Aldershot but his remains were later transferred to Brookwood Military Cemetery in Surrey.

Figure 15. Three generations of women from Walter Jones family. Standing L-R Myra Walker, nee Jones, (the child's mother); Catherine Jones, nee Gaskell, (grandmother and wife of Walter). Seated: Harriet Gaskell nee Mann, (great grandmother and Catherine's mother) holding Geoffrey Walker, born 4 March 1917.

Figure 16. This is believed to be Walter Jones with the tailor-made long wheelbarrow used to carry boxes into which the different flower tupes were placed for transport back to the packing shed circa 1910.

Figure 17. Arthur and Bess, with their three children, Keith, Muriel and Lionel.

Arthur and Bessie

In 1917, Walter's youngest brother, Arthur John Jones and his wife Bessie Edith (née Boyce) moved from their house at 145 Prospect Hill Road, Canterbury, to take over Walter's flower growing business. They were both also aged about 44 years. This change of profession was probably an opportune one for Arthur because the carriage and wheel-making business that he had been in for many years was clearly destined for extinction. (Arthur had worked as a wheelwright for Keep Bros and Wood, Importers and Manufacturers of Carriage Materials, A'Beckett St, Melbourne and later ran his own wheel making business, also in A'Beckett St.) We believe that Arthur simply carried on the flower growing business that Walter had created and did not change it in any significant way.

Arthur and Bessie brought with them their three children – Keith Erskine, Lionel Arthur, and Muriel Edith Jones - aged about 13, 11, and 8 years respectively. Like their cousins, the children also went to school at Mitcham State School and perhaps worshipped at the Mitcham Church of England Parish. After they left school in about 1919 and 1921 respectively, when they turned 14, both Keith and Lionel returned to the farm to work in their father's flower growing business. In the 1930s, they gradually took over the further development and running of the business.

Bessie's sisters and brothers, the Boyces, used to visit the farm quite regularly during the 1920s and 30s as a break from inner suburban living in Richmond. They always returned home with bundles of flowers (or "windfall" apples and pears from the neighbouring orchard) as is shown in Figures 18 and 19. Bessie was one of 11 children. Four of them married but only two – Bessie and John Clarkson Boyce – had children.

In about 1935, Arthur and Bessie were able to retire to Black Rock on Port Phillip Bay. They did however return to Tunstall in early 1942 to help on the farm (as mentioned in Lionel's letters from an Italian prison camp) and let the Black Rock house to the Furniss family who stayed there until the early 1950's. (Incidentally, Lionel was engaged to Marjorie Furniss just prior to the war, and corresponded with her for over four years, but they didn't marry on his return.)

Figure 18. Front row: Arthur, Aunt Tot, Emily Coghlan and Aunt Kath. Back row: Muriel, Keith, and Bess.

Figure 19. Muriel and her father (in cap) with three of her Boyce Uncles (Ted, Bill and Fred) on a visit to the farm in the 1930s.

Keith and Elma, and Lionel

Keith married Elma Annie Jeffery late in 1935 and they began their family, producing their three children by late in 1940. Keith and Lionel, the two brothers, continued to develop and run the business, but began to gradually take it in new directions by ceasing the production of bulbs and concentrating on the cut flower side of the business, particularly roses and orchids.

Figure 20. Keith Jones with daffodil blooms on the long wheelbarrow in the late 1920's.

This process was interrupted by the Second World War, with Lionel joining the Second 24th Battalion of the AIF in 1940 and going off to fight the Italians and the Germans in North Africa.

The next five years were extremely difficult for both the business and the families. Keith ran things virtually alone until about April 1942 when he also joined up, serving in the RAAF in East Gippsland in a training role. Muriel married Arthur Coghlan in 1942 while Lionel was in the prison camp in Italy and the couple moved to Sydney. During these years, Arthur Jones returned part-time to the farm to help. However, he was about 70 years old and this took its toll. Bessie was struggling with her health and passed away on the 15th July 1944. Arthur was unable to cope after her death and was forgetting to tend the plants so Muriel had come down from Sydney to assist but was soon to return. Keith therefore sought and was granted compassionate leave in

September 1944 to return to the farm and resume the business. Lionel returned from the war in 1945 and the business partnership with Keith resumed.

Interestingly, the second author (Lionel's daughter, Liz Ogden) holds an extensive collection of Lionel's letters home describing the journey to Egypt, his war experiences in Libya, his imprisonment in north-eastern Italy, his escape over the mountains to Switzerland from a work camp south-west of Milan, and his internment in Switzerland. The collection also includes a daily diary Lionel kept for his entire period overseas that he had managed to keep concealed from both his German and Italian captors. Lionel's diaries and his letters to Keith and Elma naturally reveal his concern about the state of the farm business and his parent's health, but they also include many observations about crops, plants, and flowers that he had seen.

Some excerpts from his diary which were written in the first few weeks after his capture at Tobruk on May 1, 1941 by Units of Rommel's Army are shown below. Incredibly, they were written at a time when he would have been dealing with the trauma of the recent battle, the disappointment of being captured, news of the death and serious injury of some of his closest mates, apprehension about what lay ahead, and inadequate food, and yet he was appreciating the flowers and plants. Quite amazing!

6/5/1941, Tuesday

Off on fatigue to Derna drome digging slit trenches - very weak on the grub. Plenty of planes in & out. Derna a pretty town - plenty of green about. Geraniums, Bougainvillea, Roses, Oleander, Lilly-Pilly's, Moreton Bay Figs, Date Palms all in plenty.

7/5/1941, Wednesday

Out on fatigue to Derna town cleaning up a Public building & washing - back to campo for dinner. Garden gay with flowers also Snaps (snapdragons) & Ursinias. Another fatigue in afternoon at Officer's quarters, beautiful building with honeysuckle & bougainvillea climbing over. Also Lantanas.

17/5/1941, Saturday

At last off to Bengazi and "What"? Hot coffee & two loaves for breakfast & off about 0800 hrs. 18 to the truck - not bad comfort - oily trucks. We were on show all day. ½ tin bully per man for lunch. Slept at barracks at Baraca for night - no tea. Camp C.O. hostile to Aussies. Nice town. Between Derna & Barce, much agriculture, good crops of barley, grapes, peaches, almonds.

Figure 21. Left: Keith and Elma Jones with daughter Helen, circa 1943. Keith had joined the RAAF and served in Victoria from 1942 to 1944. Right: Lionel in his new uniform and about to join the army, 1940.

Keith's wife, Elma, played a vital part-time role in the business for over 20 years. Apart from raising three children and preparing meals for 6 people and doing their washing, darning and ironing under fairly primitive conditions, she was the interface between the business and the retail customers in that she took the orders that the various retail florists phoned though to the farm. This often involved checking with Keith or Lionel whether particular flowers would be available, so she had a special gong made from an old brake drum that she used to bang vigorously with a steel bar (an old tyre lever) to get their attention and ask the question when they were working at a distance from the house. Remember that there were no mobile phones in those days! The question might be "Will you have a white Cattleya orchid available in 10 days-time for a wedding order we have just received?" or "Do you have any more of those small pink roses like the ones you sent last week?" She recorded all orders in a special order-book that was always kept beside the phone.

The period 1946 to about 1959 was one of considerable prosperity for the Jones flower business. During this time, Keith and Elma's children – Jeffery b 1936, Roger Keith b 1938, and Helen Rosemary Jones b 1940 - grew up on the farm and went to school at Forest Hill State School and subsequently various private secondary schools.

The children were expected to contribute to life around the farm. Tasks included chopping wood and finding kindling for the kitchen stove and hot water furnace, feeding the chickens and collecting eggs, mowing around the house, and so on. They also picked gooseberries when available and "topped and tailed" them in readiness for pie or jam making.

Lionel remained single, until aged 52 years, he married Elizabeth Joyce Brown in 1958 (Figure 22.) and built a house on a subdivision of Moore's orchard adjacent to the farm. Elizabeth Rosemary Jones (second author) was born in 1960.

This decade after the war was also one of rapid growth of Melbourne, and semi-rural regions on its outskirts such as Nunawading were enveloped by suburban development. Thus, in 1957, the Education Department purchased much of the "bottom garden" of the farm and built the Nunawading South State School. Then, in 1959 or 1960, developers purchased the "top garden", the business partnership was dissolved, and the farm's infrastructure was sold off.

Figure 22. Lionel and Bette on their wedding day. Photo Ballarat Courier

Figure 23. Roger taking orchids to the teacher at Forest Hill State School 1948.

THE "ROSEMONT" FLOWER FARM

Infrastructure

BUILDINGS, SHADEHOUSES AND WATER FACILITIES

While the 1883 advertisement of the Boldini farm (see Figures 8, and Figure A 6 in the Appendix) refers to a "4 room weatherboard cottage on the crown of the hill", it now seems certain from the rate records of the Mitcham Riding between 1899 to 1902 that this house was not actually on that portion of Boldini's farm that Walter later purchased. (Walter must have built the house (see Figures 24-26), or had it built for him and his family, between about 1901 and 1902; this resulted in a sharp increase in the rateable value of the property from 1902 onwards.)

We have no way of knowing how the Boldini farm was used between 1885 and 1901, i.e. during the period when it was owned by the Freehold Investment and Banking Co. of Australia - an investment or bankruptcy recovery company. There is evidence in the Rate Books that the property was rented by a number of families during the 1890s, so it was probably not completely neglected when Walter Jones took it over in about 1901. Similarly, we have no way of knowing whether any of Boldini's other infrastructure became incorporated into Walter's farm. The Stables/Feedshed/Garage (Figures 27 & 28), and what we referred to as the Packing Shed/Wagon Shed (Figure 29) may possibly have been.

Figure 24. The house on the "top garden" was basically a four-roomed house with a lean-to kitchen attached. We think that this photo was taken sometime in the 1920s.

Figure 25. The little-used front verandah of the house at "Rosemont", taken in about 1955. The plant on the right was a Gordonia.

Figure 26. A side view of the house circa 1956 showing the bathroom and dining room windows, the chimney of the kitchen, the Red Flowering Eucalypt (C. ficifolia) over the wash house and Lionel's "sleepout" behind the yellow-green privet hedge.

Figure 27. The Stables/ Feedshed /Garage circa 1950. The Wagon Shed is in the shade on the right.

Figure 28. The Stables/ Feedshed / Garage, the rows of Watsonias and, in the foreground, the Lorraine Lea roses. The "chook house" is in the middle right and the tennis court and the Nunawading South State School are at the rear. Photograph taken from the windmill beside the main dam in about 1959.

Figure 29. L-R: Helen and Jeffery Jones and an unknown girl in front of the Packing Shed / Wagon Shed circa 1950. The farm's Clydesdale draught horse, "Nugget", is harnessed to the farm cart.

Figure 30. Muriel, Jack, barrow and packing cases.

Figure 31. View of part of the "bottom garden" cira 1926, probably taken from the windmill/ tank stand beside the dam, showing rows of plants in the foreground and the driveway, hydrangea shade houses, and the house in the background.

Figure 32. Photograph circa 1926, probably taken from the windmill/tank stand on the "top garden", showing the Bulb Shed and various shade house structures. The house and the row of pine trees on the southern border of the "top garden" are in the distance.

LOCATION OF ASSETS

The locations of the various infrastructure assets are shown in the aerial view of the farm in 1945 (Figure 33). While a few may have carried over from Boldini's era, we believe that the majority (apart from the glasshouses) were the result of Walter Jones's efforts between about 1901/02 and 1917. We think that he was the principal builder/ developer of the farm's infrastructure and that his brother, Arthur, simply took over but did not further develop the infrastructure.

These infrastructure assets consisted of:

- The House (**No 1**);
- The Stables/Feedshed/ Garage (**No 2**);
- The Packing Shed/ Wagon Shed **(No 3**);
- a Workshop/ Woodshed/ Machinery Shed/ Playroom (**No 4** in Fig 33);
- a Bulb Shed and associated water tank (**No 5**);
- large farm Dams – in the "top" and "bottom" gardens (**Nos 6** and 7); and two Windmills, tank stands and tanks to service the large dams;
- a Shaded area for hydrangeas made by stacking brush on a timber structure (**No 8**).

Also shown are the locations of the MMBW Pipe track (**No 9),** Mt Pleasant Rd (**No 10**), the glasshouses (**No 11**), the rows of flowering plants (No **12a** and No **12b**), and Moore's apple and pear orchard (**No 13**). Preparation area for plant media (No 14).

Other infrastructure assets that are too small to show on Figure 33 include:

- two Windmills, tank stands and tanks to service the large dams;
- a Milking Shed near the main "top garden" dam;
- a large wire-netted Enclosure around Holly trees that may have also been an Aviary;
- small dams - at the south-eastern corner of the "top garden", and another even smaller one (both without windmills) near the western side of the "top garden" just beyond the Watsonias.

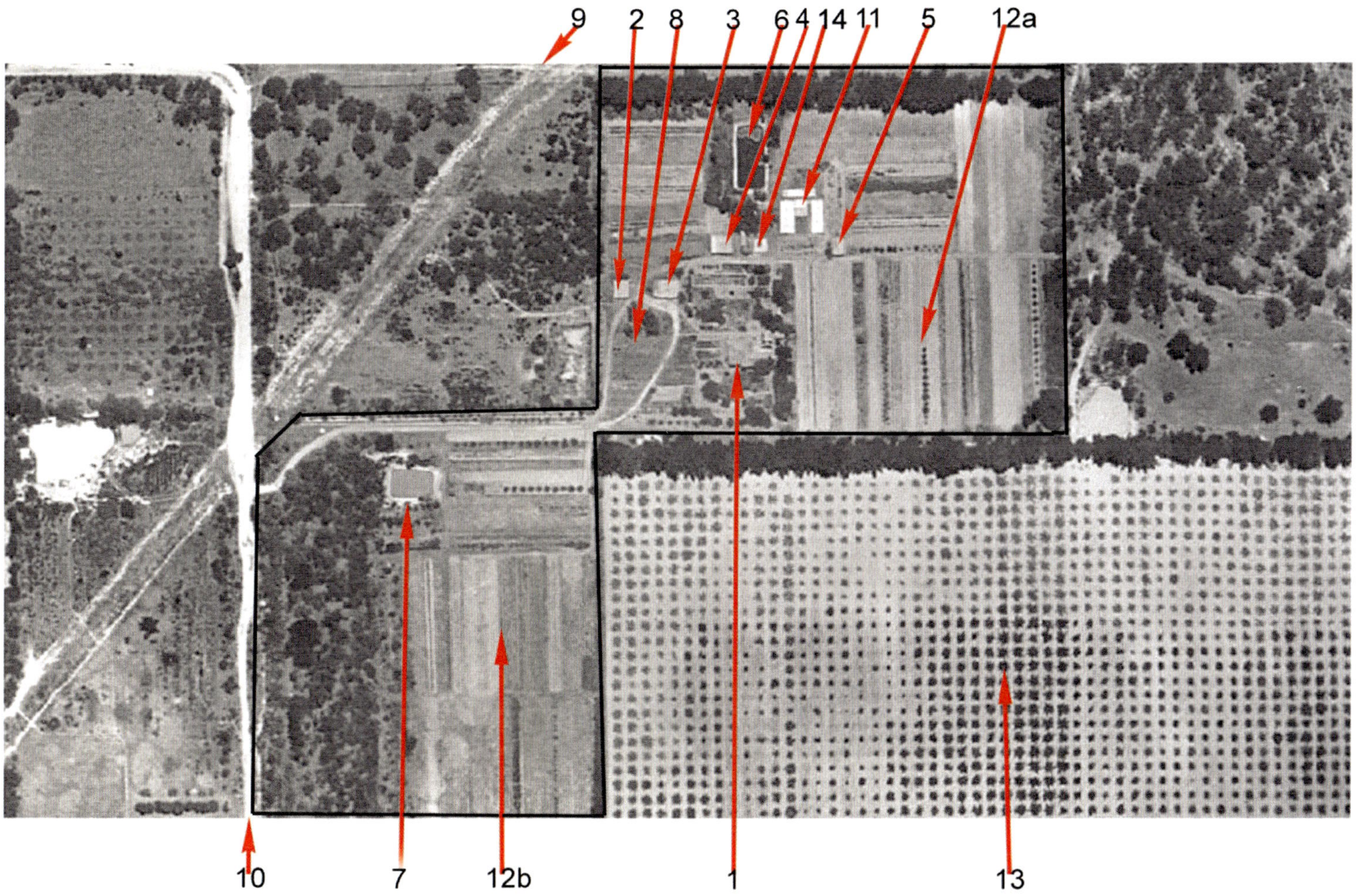

*Figure 33. 1945 Aerial Photo of the farm. Note that the entire top garden and about 60% of the bottom garden were under cultivation. Reproduced from **http://maps.whitehorse.vic.gov.au** with the permission of the Victorian Government's Department of Environment, Land, Water & Planning, and Whitehorse City Council*

Figure 34. The main farm dam with its waterlilies and the row of Pine trees (Pinus radiata) on the northern boundary of the "top garden" circa 1955. Detail of lily below.

In 1945, a single bedroom for Lionel that we referred to as "The Sleepout" was erected just prior to his return from the war. It was between the Washhouse and the entry vestibule and kitchen of the house. It was to be his bedroom for the next 13 years. (Lionel had been sent to the Convalescent Depot at Ballarat for a time on his return home in 1945. There he met and fell in love with Elizabeth Joyce Brown but it was not until 1958 that he convinced her to marry him and move to Melbourne.)

ORCHID GLASSHOUSES

Keith and Lionel Jones became interested in orchids in the late 1920s or early 1930s and began building glasshouses in which to grow them. Several glasshouses were in existence in 1940 when Lionel left for the war and at least three are visible in the aerial photograph taken in 1945 (Figure 33). Another three or four followed in the ten years after Lionel returned from the war. They were clustered close together for winter heat conservation but varied greatly in design and construction because of the Australia-wide limited availability of building materials, particularly between 1945 and 1950. The plans for one built in 1946 are shown in Figure 36. One glasshouse was specifically for Cymbidiums; it also contained a small quarantine area for orchid seedlings imported from the UK. The Cymbidium glasshouse was not heated. The plants were raised in quite large pots sitting on the ground as shown below.

Figure 35. Cymbidium Orchid glasshouse

Other glasshouses were for Cypripediums, now known as Papiopedilums (lady's slipper orchids) or for Cattleyas. The last-erected glasshouses contained Cypripediums– both the greenish-coloured and the deep red spotted one - mass-planted in beds on the bench tops rather than in individual pots. This approach represented their first attempt at mass production of orchids which could be sold more cheaply. This glasshouse was constructed like a "lean-to" along the rear end of the others and had a low roof. At the end closest to the furnace, there was a highly effective propagating bed warmed from the bottom by a copper tank containing hot water.

In about the late 1940s, a furnace was installed and hot water was circulated in steel pipes to most of the glasshouses for about four months of the year. Initially the furnace burnt coke to provide the heat but later this was changed to fuel oil. Lionel or Keith had to add more coke or check the flow of oil to the furnace at 1.00 or 2.00 AM each morning in winter. In summer, the glasshouses were whitewashed to moderate temperatures.

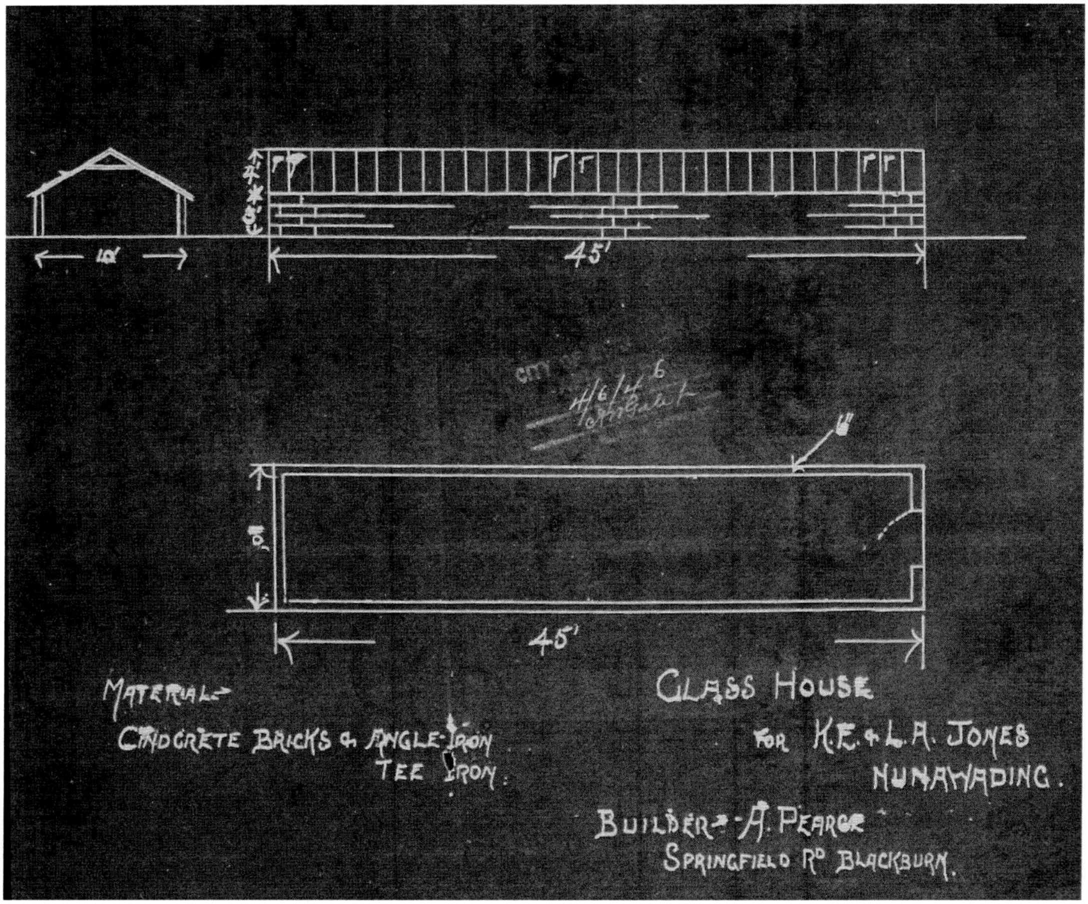

Figure 36. Plans for the first glasshouse to be built after Lionel returned from the war.

TENNIS COURT

We think that it was Lionel Jones who built a tennis court, probably during 1938/39 following Keith's and his purchase of the additional land shown in Figure 10. The court was surrounded by a well-constructed timber and wire netting enclosure and even had a shaded extension near the net for players to sit during a match. The lower part of each end of the enclosure was timbered so that a player could practice shots by hitting balls against this wall. The surface of the court was made of compacted earth.

Farm Animals

HORSES

Lighter and shorter-distance transport of goods, e.g. to Tunstall Station and within the farm, was provided by a 2-wheeled cart. In the 1920s and 1930s this was presumably pulled by "Nell" (Figure 37) or the lighter horse "Jack" (Figures 38 & 39), but in the 1940s and 1950s these horses were replaced by "Nugget" (Figure 37). Nell and Nugget were heavy draught horses and were used for cultivation the fields. Arthur Jones and family had a horse-trap for people transport in the 1920s and 30s until it was replaced by a late-1920s Dodge or Chevrolet sedan with a canvas hood.

Figure 37. Top, "Nugget", with Muriel' son, John Coghlan, astride. Also shown are the family's Airedale, "Ruggles", and an unknown black and white dog. Bottom: "Nell" and Arthur's daughter, Muriel.

Figure 38. Light draught horse "Jack" and the farm's cart. Jack worked on the farm and provided transport and draught in the 1920s and 1930s.

Figure 39. Arthur Jones and his daughter Muriel in the horse-trap drawn by Jack, circa 1936

MISCELLANEOUS OTHER ANIMALS

The farm had a milking cow – usually of Jersey breed - probably from the outset in the early 1900s. This practice continued until the late 1950s. In the 1950s there were also a dozen or so chickens or "chooks"; an Airedale dog, "Ruggles"; a grey cat, "Smokey"; and for a time, ducks being raised by Roger for pocket money.

Figure 40. The farm's Jersey cow standing in the bottom of the main farm dam during a major drought – perhaps in 1932 - when the dam dried up completely and the goldfish had to be rescued. Chooks below.

Figure 41. Helen, Roger, and Jeff on the running board of the old Dodge or Chevrolet.

In about 1949, both Lionel and Keith Jones, by then more prosperous, purchased brand new cars – an MG 4-door sedan and an Austin 16 sedan respectively.

Land Preparation

The draught horses were also used for cultivation – pulling either a single mouldboard plough or a set of harrows. The farm had a sled for transporting heavy items around the farm. A walk-behind Howard Gem Rotary Hoe similar to that shown in Figure 42 satisfied the final cultivation needs from the 1930s through to the late 1950s, providing an excellent fine seedbed for planting. In the early 1950s, in the days before herbicides were invented, a Turner 2-wheeled brush cutter was also purchased and used extensively for controlling grass and weed growth.

Figure 42. The farm had a Howard Rotary Hoe similar to this one for its finer cultivation needs from the 1930s through to the late 1950s. Image courtesy of Graces Guide to British Industrial History.

Plant Resources

PERENNIAL PLANTS

The extensive plantings of perennial trees and shrubs that we believe were made on the farm by Walter Jones between 1901 and 1917 can also be regarded as a form of infrastructure. They were extremely important as they provided a regular annual supply of flowers and foliage, and the revenue therefrom, for over 50 years and at minimal annual cost. Many of the species involved are listed in Table 1 and depicted in the following photographs.

Figure 43. Lt. WA Flowering Gum; Rt. Rhododendron.

Figure 44. A row of white Agapanthus, circa late 1920s.

Figure 45. Daffodils and roses growing in the "top garden" circa late 1920s.

Figure 46. Dahlias and roses growing on the western side of the "top garden" in the late 1950s.

Figure 47. Hydrangeas under a shade house made from local materials, circa 1950.

Figure 48. Flowering plums taken in the late 1950s.

TABLE 1: THE PRINCIPAL PERENNIAL PLANTS GROWN ON THE FLOWER FARM, CIRCA 1910 – 1959*	
Common Name	**Genus, Species, and Comments**
Arum Lily	*Zantedeschia aethiopica,* the white funeral flower. The farm had one large clump in an area waterlogged by drainage from the bathroom.
Azalea	*Azalea or Rhododendron spp,* within the family Ericaceae. Azaleas are shade tolerant and bloom in spring, their flowers often lasting several weeks. In the 1950s, the farm had a number of bushes housed in small shade house near the Bulb Shed.
Berberis	*Berberis thunbergii* (Japanese barberry, Thunberg's barberry, or red barberry) is native to Japan and eastern Asia, though widely naturalized in China and North America. It is a dense, deciduous, spiny shrub which grows 0.6 to 2.5 m high. The farm had a large thicket approx. 20 m2 in area beside the brick path from the house to the glasshouses. Its attractive red foliage and berries were sold to florists.
Bird of Paradise	*Strelitzia reginae,* perennial plants native to South Africa. Also known as the Crane Flower. The farm had several clumps near the house and the flowers and leaves were sold for use in large floral displays.
Begonia	*Begonia* is a genus of perennial flowering plants in the family *Begoniaceae.* The genus contains more than 1,800 plant species that are native to moist subtropical and tropical climates. Some species (eg the *Rex begonias*) are commonly grown indoors as ornamental houseplants. Multicoloured *Rex bogonias* thrived when grown under the benches in the Cattleya glasshouse.
Bugle Lily	*Watsonia,* a genus of plants in the iris family, subfamily *Crocoideae,* native to southern Africa. The most commonly cultivated species is the pink-flowered *Watsonia borbonica* and its white mutant 'Arderne's White'. These were crossed with *Watsonia meriana* and other species in the early 20th century. The farm had one large field below the main dam devoted solely to Watsonias.
Camellia	*C. japonica and C. sasanqua and their hybrids.* The farm had numerous cultivars with red, white, and pink flowers. One very old clump of white flowered Camellias near the front verandah of the house was about 30 ft high.
Cotoneaster	*Cotoneaster spp.* There was a medium-sized tree beside the driveway that was harvested for its red berries and attractive foliage.
Crab Apple	*Malus spp.* A compact ornamental tree, providing blossom in Spring and colourful fruit in Autumn that often persist throughout Winter. Numerous hybrid cultivars have been selected. The farm had several trees beside the elliptical drive at the end of the brick entrance path.

Daphne	*Daphne odora,* a species of flowering plant in the family *Thymelaeaceae,* native to China, later spreading to Japan and Korea. It is an evergreen shrub, grown for its very fragrant, fleshy, pale-pink, tubular flowers, each with 4 spreading lobes, and for its glossy foliage. The farm had several bushes and the flowers were sold.
English Hawthorn	*Crataegus monogyna.* The top farm had a long row of trees, harvested for their flowers and the attractive red berries.
English Lavender	*Lavendula* is a genus of 47 known species of flowering plants in the mint family, *Lamiaceae.* They are annual or short-lived herbaceous perennial plants, or shrub-like perennials, native to the Old World from Cape Verde and the Canary Islands, Europe across to northern and eastern Africa, the Mediterranean, SW Asia to SE India. Many are cultivated in temperate climates as ornamental plants, as culinary herbs, and commercially for the extraction of essential oils. The most widely cultivated species, *Lavandula angustifolia,* is English or French lavender. In most species, the leaves are covered in fine hairs or indumentum, which normally contain the essential oils. The "top garden" had several rows of Lavender that were harvested for their flowers or dried for their scent.
European Holly	*Ilex aquifolium,* a small, evergreen tree with red berries. The farm had two large trees (5 m) in the wired enclosure to the south of the house.
Forget-me-not	*Myosotis* is a 74 species genus of flowering plants in the family *Boraginaceae.* In the northern hemisphere they are colloquially denominated forget-me-nots or Scorpion grasses. The genus is largely restricted to western Eurasia with circa 60 confirmed species and New Zealand with circa 40 confirmed species. The species is common throughout temperate latitudes because of the introduction of cultivars that are popular in horticulture. They prefer moist habitats. In locales where they are not native, they frequently escape to wetlands and riverbanks.
Fuschia	*Fuschia spp.* Almost 110 species of Fuchsia are recognized; the vast majority are native to tropical or subtropical South America, but a few occur north in Central America to Mexico. Most fuchsias are shrubs from 0.2 to 4 m tall, usually with leaves that have serrated margins. They can be either deciduous or evergreen, depending on the species. The flowers have a pendulous "teardrop" shape and are displayed in profusion throughout the summer and autumn. The sepals are bright red and the petals purple, but the colours can vary from white to dark red, purple-blue, and orange. The top garden of the farm had a number of plants.
Gardenia	*Gardenia,* a genus of flowering plants in the coffee family, *Rubiaceae,* native to the tropical and subtropical regions of Africa, Asia, Madagascar and Pacific Islands. They are evergreen shrubs and small trees growing to 15 m tall. Leaves are dark green and glossy with a leathery texture. The flowers are solitary or in small clusters, white, or pale yellow. Many species are strongly scented. The top garden had a number of plants harvested for their flowers and scent.

Gooseberry	*Ribes uva-crispa. synonym Ribes grossularia* or Gooseberry is a species of *Ribes* which also includes the currants. It is native to Europe, the Caucasus and northern Africa. Gooseberry bushes produce an edible fruit. "Gooseberry bush" was 19th-century slang for pubic hair, and from this comes the saying that babies are "born under a gooseberry bush". The gooseberry is a straggling bush growing to 1.5 m in height and width, the branches being thickly set with sharp spines. The fruit are berries, usually green, but there are red (to purple), yellow, and white variants.
Gordonia	*Gordonia* is a genus of flowering plants in the family *Theaceae,* related to *Franklinia, Camellia and Stewartia.* Of the roughly 40 species, all but two are native to southeast Asia in southern China, Taiwan and Indochina. They are evergreen trees, growing to 10–20 m tall. The leaves are alternately arranged, simple, serrated, thick, leathery, glossy, and 6–18 cm long. The camelia-like flowers are large and conspicuous, with mostly 5 white petals. The farm had several plants close to the front verandah of the house. The stiff glossy leaves were also used in wreaths.
Guava	Probably Strawberry guava, *Psidium littorale.* One tree with small purple fruit was located in the Holly enclosure south of the house.
Heather	*Calluna vulgaris* is the sole species in the genus in the flowering plant family Ericaceae. It is a perennial shrub growing to a maximum height of 1 m in Europe and Asia Minor growing on acidic soils in open sunny or moderately shady situations. Clumps of heather growing on the old Boyle block to the west of the tennis court were occasionally cut for their flowers.
Hellebores	The Eurasian *genus Helleborus* consists of approx. 20 species of herbaceous or evergreen perennial flowering plants in the *family Ranunculaceae.* Despite names such as "winter rose", "Christmas rose" and "Lenten rose", hellebores are not closely related to the rose. Many hellebore species are poisonous. The often greenish-coloured flowers have five petal-like sepals surrounding a ring of small, cup-like nectaries which are actually "petals" modified to hold nectar. The sepals do not fall as petals would, but remain on the plant, sometimes for many months. The farm had plants near the Azalea house.
Hibiscus	*Hibiscus,* a genus of flowering plants in the mallow family, *Malvaceae,* native to warm-temperate, subtropical and tropical regions. Cultivars are renowned for their large, showy flowers. The farm had several plants growing beside and just west of the house.
Indian Hawthorn	*Rhaphiolepis indica,* the Indian hawthorn, India hawthorn or Hong Kong hawthorn, is an evergreen shrub in the *family Rosaceae.* The species is native to an area from southern China, Japan, Laos, Cambodia, Thailand and Vietnam. It is grown for its decorative pink flowers and stiff branches and leaves. The farm had a number of bushes for both foliage and flowers.

Hydrangea	*Hydrangea,* a genus of 70–75 species of flowering plants native to southern and eastern Asia (China, Japan, Korea, the Himalayas, and Indonesia). Most are shrubs 1 to 3 meters tall are all deciduous. The farm had large plantings of pink and blue hydrangeas beneath shade shelters within and beside the elliptical area enclosed by the driveway.
Japonica	*Chaenomeles japonica* or flowering quince, a relative of the edible quince, is in the *Rosaceae or rose* family and originated from China and Japan. Many people still know it as the flowering japonica. The mostly single flowers appear on bare stems from early winter till early spring, when the foliage reappears. The flowers are mainly pinks and reds, with a cluster of golden yellow stamens in the centre. They produce a small quince-like fruit that can be used to make a jelly or paste. The farm had several very large Japonica thickets >50 square meters in area with access tracks cut out within them. They were harvested every winter.
Lilac	*Syringa vulgaris,* a species of flowering plant in the olive family *Oleaceae,* native to the Balkan Peninsula, where it grows on rocky hills. The farm had a number of trees that were valued for their lilac or white flowers and beautiful scent.
Magnolia	*Magnolia spp.* The farm had several trees near the Packing Shed; they had large white and pinkish flowers.
Maple	*Acer spp.* Most maples are deciduous trees growing to about 10–45 m, and many are renowned for their autumn leaf colour. The farm had 2 large trees just outside the back bedroom. We don't think that they were used commercially.
May Bush	*Spiraea cantoniensis* is a spectacular white flowering bush to a height of about 3 m that originated in SE China. It flowers is May in the northern hemisphere, hence its name, but in September- October in the southern hemisphere. The "top garden" had several bushes just to the north of the house that were cut for their flowers.
Nasturtium	*Tropaeolum* commonly known as nasturtium is a genus of roughly 80 species of annual and perennial herbaceous flowering plants - the only genus in the family Tropaeolaceae. Nasturtiums produce an oil similar to that of watercress (Nasturtium officinale). The genus, native to South and Central America, includes several very popular garden plants that have showy, often intensely bright flowers, and rounded, shield-shaped leaves with the petiole in the centre. The farm did not make any commercial use of this plant.
Nymphaea	*Nymphaea* or water lily is a genus in the *family Nymphaeaceae.* Many species are cultivated as ornamental plants, and many cultivars have been bred. Water lilies are rhizomatous perennial herbs. The leaves grow from the rhizome on long petioles and usually float on the water surface. The flowers rise out of the water or float on the surface. The "top farm" dam had several clumps with red, white, and blue flowers. They were harvested from a small row-boat using a long wire that slid down the flower stem.

Persimmon	*Genus Diospyros, family Ebenaceae.* The most widely cultivated species is the Oriental or Japanese persimmon, *Diospyros kaki* that is native to Japan, China, Korea, Burma and Nepal. It is deciduous, with broad, stiff leaves that have attractive autumn colours. The fruits have a soft to occasionally fibrous texture but a high tannin content, making the unripe fruit astringent and bitter but delicious when fully ripe. The farm had 2 or 3 trees on the lawn area to the north of the house that were spectacular in winter with their large orange fruits on otherwise bare trees. The only commercial value the trees provided was from their autumn leaves.
Prunus	*Prunus spp.* A genus of trees and shrubs, which includes the plums, cherries, peaches, nectarines, apricots and almonds. *Prunus* species are spread throughout the northern temperate regions and many are cultivated for fruit and ornament. The farm had numerous pink or white flowering trees, also some species with reddish-purple foliage growing in various parts of both gardens. The pink blossoms were intensively harvested in late winter and early spring.
Pussy Willow	*Salix spp,* a relative of willows and sallows. In spring, it bears attractive furry catkins. The farm had one large tree near the Packing Shed that was harvested for use in large flower arrangements.
Quince	*Cydonia oblonga,* the sole member of the genus *Cydoni,* family *Rosaceae* is a small deciduous tree that bears a large pome fruit, like a pear that is bright golden-yellow when mature. There was one tree on the driveway. Its fruit were occasionally used in the house.
Red Gum Flowering	*Eucalyptus* now *Corymbia ficifolia,* the Western Australian flowering gum. There was one huge tree about 7 metres high beside the Wash House.
Red Hot Poker	*Kniphofia ensifolia, family Asphodelaceae,* is native to Africa. A perennial species with broad, strap-shaped foliage up to 1.5 m long. Plants produce spikes of upright, bright red, orange and yellow flowers above the foliage. The farm had at least one large clump near the front gate.
Rhododen dron	*Rhododendron spp.* A genus of 1,024 species of woody plants in the heath family (*Ericaceae*), either evergreen or deciduous, and found mainly in Asia, although it is also widespread throughout the Southern Highlands of the Appalachian Mountains of North America. Plants have brightly coloured flowers produced from late winter through to early summer. The "top garden" of the farm had many bushes near the Bulb Shed and in other places.
Rhus	*Toxicodendron succedaneum, previously Rhus succedanea* - a highly toxic, allergy-causing tree with attractive red foliage. There were several plants in the "top garden".

Rose	*Rosa spp* are woody perennial plants of the genus *Rosa,* family *Rosaceae* with >100 species and thousands of cultivars. They can be erect shrubs, climbing or trailing with stems that are often armed with sharp prickles. Flowers vary in size and shape and are usually large and showy, in colours ranging from white through yellows and reds. Most species are native to Asia, with smaller numbers native to Europe, North America, and northwestern Africa. Both the "top" and "bottom" gardens had extensive plantings of old roses that were bred between 1880 and 1940. The main cultivars were Black Boy, Mlle Cecile Brunner, Lady Hillingdon, Lorraine Lea, Madame Rivers, Papa Gontier, Peace, and Sunny South.
Stephanotis	*Stephanotis floribunda syn. S. jasminoides* (Madagascar jasmine, waxflower, Hawaiian wedding flower, bridal wreath) is a species of flowering plant in the family *Apocynaceae,* native to Madagascar. Growing to 6 m or more, it is an evergreen woody climber with glossy, leathery oval leaves and clusters of pure white, waxy, intensely fragrant tubular flowers. The farm sold the fragrant flowers from several plants.
Tecoma	*Tecoma capensis* or Cape Honeysuckle, an erect, scrambling shrub that grows to 2–3 m in height and width. The flowers are tubular, narrow, about 7.5 cm long, in colours ranging from orange to orange-red to apricot. There was a large scrambling plant near the back door of the house with bright orangey-red flowers. The flowers were very attractive to small black ants. This plant was of no commercial value to the farm.
Tulip Tree	*Liriodendron tulipifera,* or American tulip tree, native to eastern North America from Southern Ontario and Illinois eastward to SE Massachusetts and Rhode Island, and south to central Florida and Louisiana. It can grow to more than 50 m. The farm had only one large tree with lovely large pale greenish-yellow flowers.
Violets	*Viola odorata* is a species of the *genus Viola,* native to Europe and Asia. It is a hardy herbaceous perennial with highly aromatic dark purple or white flowers. The leaves and flowers are all in a basal rosette and the plant spreads with stolons (above-ground shoots). The sweet scent of this flower has proved popular, and has consequently been used in the production of many cosmetic fragrances and perfumes. The scent is very distinctive with only a few other flowers having a remotely similar odor. Violets were grown commercially on the farm in the 1920s and 30s but production ceased because of the time and back-breaking labour involved in picking them.

Plant descriptions were abbreviated from entries in Wikipedia. The **common names of plant types that were of particular commercial significance are shown in Bold.*

Courtesy of creativecommons.com.
Source: https://www.flickr.com/photos/47523448@N008232440

Figure 49. Some of the various colours of daffodils shown here. Photo credit Matthew Jones.

PLANTS WITH BULBS, CORMS, TUBERS OR RHIZOMES

In the 1920s and early 1930s, and perhaps much earlier, Walter Jones's business seems to have had a strong reliance on the production and sale of bulbs to major Victorian Seed and Plant Merchants as well as to retail outlets. In fact, bulbs were such an important part of the business that there was a large "Bulb Shed" (Figure 32) approximately 6 m x 15 m for sorting, drying, storing and packing bulbs. The main trade was in various cultivars of Daffodils, but smaller numbers of other bulbous or cormous-plants were also grown, either for their bulbs or for cut flowers. The main bulbs grown on the property are listed in Table 2. Clients for the supply of bulbs in orders ranging from 50 to 1000 bulbs included the individuals and companies shown in Table 3. Appendix 3 explains the difference between bulbs, corms, tubers and rhizomes.

TABLE 2: PLANTS WITH CORMS, BULBS, TUBERS OR RHIZOMES

Common Names	**Species and Comments**
Daffodil	*Narcissus* is a genus of predominantly spring perennial plants of the *Amaryllidaceae* family that are commonly known as daffodil or jonquil. *Narcissus* has conspicuous flowers with six petal-like tepals surmounted by a cup or trumpet-shaped corona. The flowers are generally white or yellow with either uniform or contrasting coloured tepals and corona. A wide variety of species and cultivars were grown, including the following Daffodil cultivars: Albatross, Argent, Bath's Flame, Blackwell, Campernelle, Cardinal, Cervantes, CJ Backhouse, Cloth of Gold, Commodore, Cynosure, Double Roman, Duke of Bedford, Early Bird, Emperor, Empress, Fairy Queen, Flitter, Grand Monarch, Golden Bell, Golden Eagle, Gwyther, Harold Finn, King Alfred, Lemon Drop, Leonie, Lucifer, Lulworth, Madam de Graffe, Madge Matthew, March White, Maximus, Moonbeam, MJ Berkeley, Mrs Langtree, Mrs Robert Sydenham, Mrs Thompson, Nestalla, Olympia, Paper White, PR Barr, President, Princeps, Queen Bess, Queen of the East, Seagull, Seville, Sunrise, Soleil d'Or, Southern Gem, Sparkler, Torch, Victoria, War Flame, Waterwitch, White Pearl and White Poet. Members of the Narcissus genus were the principal product of the farm from about 1910 to 1930 as both flowers and bulbs. They were grown extensively and Walter exhibited them at meetings of the Canterbury Horticultural Society.
Dahlia	*Dahlia* is a genus of bushy, tuberous, herbaceous perennial plants native to Mexico. A member of the *Asteraceae* (or *Compositae*). Related species include the sunflower, daisy, chrysanthemum, and zinnia. There are 42 species of dahlia, with hybrids commonly grown as garden plants. Flower forms are highly variable, with one head per stem 5-30 cm in diameter. The stems are leafy, ranging in height from 30 cm to 2.4 m. Like most plants that do not attract pollinating insects through scent, they are brightly colored, displaying most hues, with the exception of blue.

Gladiolus	*Gladiolus* (from Latin, the diminutive of gladius, a sword) is a genus of perennial cormous flowering plants in the *iris family (Iridaceae)*. The genus occurs in Asia, Mediterranean Europe, South Africa, and tropical Africa. Gladioli grow from rounded, symmetrical corms that are enveloped in several layers of brownish, fibrous tunics. The spectacular giant flower spikes in commerce are the products of centuries of hybridisation and selection. The flower spikes are large and one-sided. These flowers are variously coloured, ranging from white and cream to pink, red and light purple. They can make very good cut flowers for display. Gladioli were grown regularly on the farm and corms were stored in the Bulb Shed.
Lilies	Both gardens had a wide variety of lilies. The following genera and species are mentioned at various times in the farm records: *Amaryllis, (e.g. A. belladonna, Belladonna Lily); Habranthus, Rain Lily; Hippeastrum; Lilium (e.g. L. henryii, Henry's Lily; L. longiflorum, Easter lily); Roman Hyacinth (Hyacinthus orientalis); Ixias, Ixia spp); and Canna lilies (Family Cannaceae).*
Lily of the Valley	*Convallaria* is a sweetly scented, highly poisonous woodland flowering plant that is native throughout the cool temperate Northern Hemisphere in Asia and Europe. Other names include May bells, Our Lady's tears, and Mary's tears. It is possibly the only species in the genus *Convallaria* in the family *Asparagaceae*. The plant is an herbaceous perennial that forms extensive colonies by spreading underground stems called rhizomes. New upright shoots are formed at the ends of stolons in summer; these grow in the spring into new leafy shoots that still remain connected underground to the other shoots, often forming extensive colonies. The farm had several patches of Lily of the Valley that produced fragrant flowers for sale.
Scillas Bluebells	*Scilla* is a genus of about 50 to 80 bulb-forming perennial herbs in the family *Asparagaceae*, subfamily *Scilloideae*, native to woodlands, subalpine meadows, and seashores throughout Europe, Africa and the Middle-East. A few species are also naturalized in Australia, New Zealand and North America. Their flowers are usually blue, but white, pink, and purple types are known; most flower in early spring, but a few are autumn-flowering.
Snowdrops Snowflakes	*Galanthus* is a small genus of about 20 species of bulbous perennial herbaceous plants in the family *Amaryllidaceae*. The plants have two linear leaves and a single small white drooping bell shaped flower with six petal-like tepals in two whorls. The smaller inner petals have green markings. Most species flower in winter or early spring. Snowdrops are sometimes confused with the two related genera within the tribe *Galantheae*, snowflakes *Leucojum* (Snowflakes) and *Acis*.
Spanish Iris	An *iris* native to Spain and Portugal and also found in Corsica, South West France, southern Italy, Algeria and Tunisia. It is bulbous and bears blue, violet, white or yellow flowers around 6 cm wide. The plant can reach up to 80 cm before flowering. The greyish-green leaves die after the flowers fade. Spanish Iris is favoured by florists for its striking colour combinations.
Valottas	*Cyrtanthus elatus* or the Scarborough Lily has large, vivid scarlet, lily-like flowers that tower above the strap shaped leaves. Growing from bulbs, they produce these very showy flowers from mid-Summer through to Autumn, then the foliage dies back.

The note and invoice from Walter Jones to his brother in 1931 (Figure 50) indicates that Walter maintained his interest in Daffodils for many years after he left the farm. Given the number of bulbs noted in the invoice, it also suggests that daffodil production and sale was still an important part of Arthur's flower growing business in 1931.

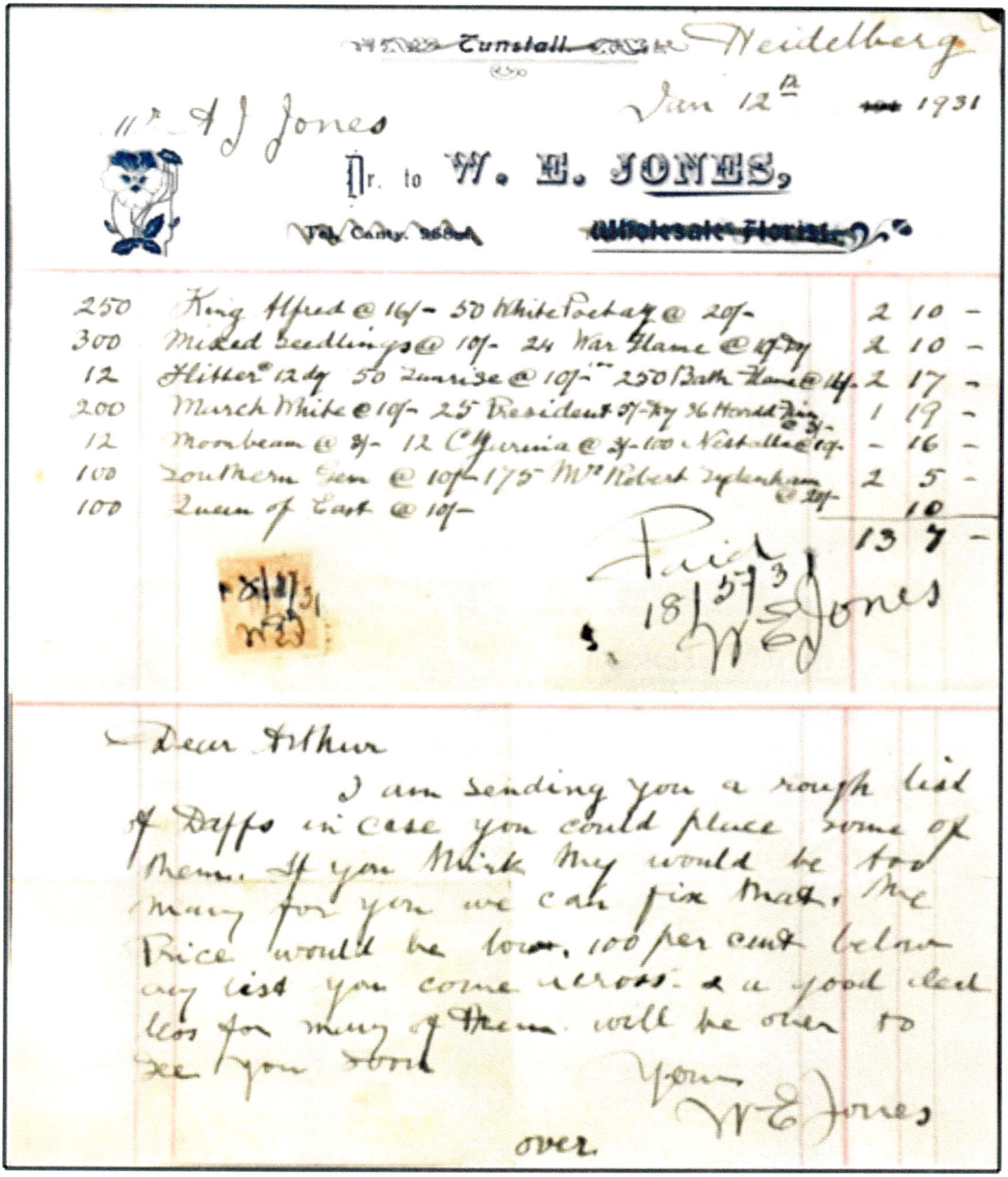

~~Tunstall~~ Heidelberg

Jan 12th 1931

Mr A J Jones

Dr. to W. E. JONES,

Tel. Canty. [illegible] Wholesale Florist.

250	King Alfred @ 16/- 50 White Poetaz @ 20/-	2 10 -
300	Mixed Seedlings @ 10/- 24 War Flame @ [illegible] doz	2 10 -
12	Glitter @ 12 doz 50 Sunrise @ 10/- 250 Bath Flame @ 14/-	2 17 -
200	March White @ 10/- 25 President 5/- doz 36 Herald [illegible] @ 3/-	1 19 -
12	Moonbeam @ 3/- 12 C[illegible] @ 3/- 100 [illegible] @ 1/-	- 16 -
100	Southern Gem @ 10/- 175 Mrs Robert Sydenham @ 2/-	2 5 -
100	Queen of East @ 10/-	10
		13 7 -

Paid 18/5/31 W E Jones

Dear Arthur

I am sending you a rough list of Daffs in case you could place some of them. If you think they would be too many for you we can fix that. The price would be low, 100 per cent below any list you come across & a good deal less for many of them. Will be over to see you soon

Yours W E Jones

over.

Figure 50. 1931 invoice from Walter Jones to his brother Arthur.

ORCHIDS

The farm had extensive collections of orchids, mainly in the Cymbidium, Cypripedium or Paphiopedilum (lady slipper) and Cattleya genera. Each was allocated a glasshouse because of their differing environmental requirements. The most demanding were the Cattleyas that originated mainly in Central and South America. The Cattleya glasshouse was therefore maintained at a higher temperature and humidity than the others, with the humidity being achieved by keeping the gravel/ ash floor moist at all times. The least demanding were the Cymbidiums. Keith and Lionel achieved excellent Cattleya growth using a liquid fertiliser consisting of diluted effluent from the septic tank of the house.

Both Lionel and Keith were long-established members of the Victorian Orchid Society, so were keen to exhibit their best blooms at monthly meetings and at the Annual Orchid Shows that were held in the Melbourne Town Hall or in the showrooms of Preston Motors (Corner of Russell and Little Collins Streets) in the city. Like many other of the society members, they were orchid aficionados, winning many prizes and awards, particularly for their Cypripediums and Cattleyas where the shape and form of the flowers was of great importance. However, they were also medium to large-scale growers of orchids for the cut-flower trade where these factors were of lesser importance.

Lionel continued his interest in orchids into the 1960s and 1970s and became the President of the Victorian Orchid Society, an orchid judge, and ultimately a life member. He and Keith were also founding members of the Melbourne Eastern Orchid Society which was formed in 1949.

Figure 51. Slipper orchids of the type grown in the glasshouses on "Rosemont".

Figure 52. White Cattleya orchids. The one on the right was named "Mrs KE Jones" in 1936-37 after Keith's wife, Elma

Figure 53. The Jones stand at a Victorian Orchid Society Yearly Show.

Labour

Nothing is known about Walter Jones' use of casual labour on the farm but, according to notes in the Farm Ledger Book, Arthur Jones employed substantial casual labour at daily or weekly rates in the period 1918 to 1932, probably continuing the practice used by his brother. Casual labourers named Thomas Galvin, Jack Merton, Jack or John Schafer, Henry West, Mr Stafford, Cliff Dows and Edward Munt were employed more regularly and for longer periods than the others listed. Arthur's requirement for casual labour, however, declined towards the end of this period, presumably because his sons, Keith and Lionel Jones were then also working in the business and beginning to assume more responsibility. From the mid-1930s right through to 1960 there is no evidence that any labour was employed by Lionel and Keith.

54. Employee Harry West weeding daffodils in the "bottom garden", circa early 1920s.

TABLE 3: CASUAL STAFF EMPLOYED BY AJ JONES AND SONS 1918 TO 1932		
Year	**Name**	**Address**
1918	CARMODY, Dan	
	COLLINS, Pat	
	GALVIN, Thomas	
	GOLDSMITH, W	
	HULL, J	
	MATTERS, Richard	Mitcham Rd
	McCONNELL, G	
	McLEAN, Hugh Colin	
	MERTON, Jack	
	SCHAFER, Jack	
	WEST, H	
1919	BENTLEY, James	Whitehorse Rd, Mitcham
	COCKRANE, Percy	
	CORDELL, A	
	CRAFT, William	Whitehorse Rd
	LEWIS, Mr	Mitcham
	MALONEY, Edward	
	PITMAN, James	Whitehorse Rd
	SCHAFER, John	Canterbury Rd, Mt Pleasant
	THOMPSON	
	WEST, Henry	Whitehorse Rd, Mitcham
1920	CRACKER, Henry	
	EDWARDS, Alfred	East Burwood
	JACOBS, Jack	Springvale Rd, Tunstall
	MARSH, Daniel	
	MERTON, Miss	Whitehorse Rd, Tunstall
	NEUMANN, William	Surrey Hills
	STAFFORD, Mr	Kent Rd, Box Hill & Whitehorse Rd, Mitcham
	WATSON, Walter	Whitehorse Rd, Tunstall & 6 Seymour Rd, Elsternwick
1921	BALL, Clarence	Station St, Box Hill
	WEST, Henry	Whitehorse Rd, Mitcham
1922	JACKSON, Arthur	Ringwood
	JOHNSON, Tom	Tunstall
	O'LEARY, G	Tunstall
	PITMAN, Victor	
	THORNELL, J	
	WARREN, L	
	WEST, Henry	Whitehorse Rd, Mitcham
1923	**WEST, Henry**	Whitehorse Rd, Mitcham

1924	MILLER, Roy	Bedford Rd, Ringwood
	WEST, Henry	Whitehorse Rd, Mitcham
1925	**DOWS, Cliff**	Whitehorse Rd, Box Hill
1926	BAUDENETTE, Victor	Main St, Blackburn
	HAMIL, Henry	Oakleigh
1927	HALL, HG	
	MUNT, Edward	Mt Pleasant Rd & Springvale Rd
1928	JONES, John	Station St Tunstall/ Box Hill
1929	CALTON, William	Whitehorse Rd, Box Hill
	PILVEN, R	Tunstall

Figure 55. Keith in his early 20s at about 1926, with an armful of daffodil blooms.

Transporting Flowers and Bulbs to Markets

Walter Jones had a four-wheeled heavy wagon like the one illustrated below that he presumably used to transport his produce to market. He may have sold sufficient produce at either the Kew Market or the Queen Victoria Markets in Melbourne to require him to have a heavy-duty vehicle. Arthur Jones may have also used the wagon into the 1930s but after that it sat unused for many years in the wagon shed until it was disposed of in the late 1940s.

Figure 56. WE Jones had a wagon somewhat similar to the one in this photograph entitled "Horse-drawn wagon on Hessels Road, Harkaway, Victoria, 1899-1914". Taken by Michael J Drew, 1873-1943, photographer, it is out of Copyright. Refer http://handle.slv.vic.gov.au/10381/246705.

Family lore talks of an early dispute between Walter Jones and a neighbour named Boyle over access to the Lilydale Railway line and/ or Whitehorse Rd. The 1892 map (Figure 4 and Figure A 2 in the Appendix) shows an easement or road running southwards from White Horse Road and the railway line to the northern edge of Allotment 116 G, between properties owned by the Tunstall Brick Company and J. McClelland (i.e. down the middle of the original Allotment 118 B). However, getting access to this from the Jones Flower Farm would have required passage across a small portion of Joseph Boyle's land. This is presumably the origin of the dispute. Further subdivision of Allotments 117 and 118, particularly the blocks owned in 1892 by Mrs. Samuels and the Tunstall Brick Co, and the creation of easements which allowed the completion of Mt Pleasant Rd – i.e. from Canterbury Rd right through to the railway line - presumably solved the problem eventually.

Figure 57. Tunstall Railway Station circa 1920.

Tunstall Station (Figure 57) had a woman in charge from at least 1904 until 1917 and probably considerably longer – the consequence of which was that it did not handle "goods". This certainly applied to the receipt of "goods" from other locations and presumably applied also to the dispatch of "goods" from this station. The Nunawading Council protested about this situation in 1904 as follows:

The Age (Melbourne, Vic.), Wednesday 7 December 1904, page 8

TUNSTALL STATION. At the ordinary meeting of the Nunawading council last night Cr. Husband moved that a letter be sent to the Commissioners of Railways protesting against their decision to place a woman in charge of the Tunstall station. The motion was seconded by Cr. Rooks and carried.

Similar protests and deputations to the Railways Commissioners continued at intervals between then and 1909 when they ceased because the Commissioners insisted that the volume of traffic did not justify a change. The next we hear about Tunstall in the newspapers of the times (apart from level crossing accidents, etc.) are requests and delegations about the need for a new Station building at Tunstall in 1927 and 1938. This new station presumably would have been of sufficient status to be staffed by a man. However, it would appear that a new station may not have been provided until sometime in the early-mid 1940s.

In view of the non-availability of Tunstall for "goods", it is likely that for much of Walter Jones' time at Tunstall, and probably part of Arthur Jones' time, they would have used Mitcham Station for the dispatch of the farm's produce. In this case, it might have been more convenient for them to access the lane running down the middle of Allotment 118 A (between blocks owned by the Metropolitan Brick Company and Scott). This is probably the future Rooks Rd (see Figure 3 and particularly Figure A 2 in the Appendix). Accessing it from the "top garden" would have only required passage across the unoccupied block 116 F. (In the late 1940's, there was a wire gate in about the middle of the eastern side of the "top garden" that would have provided such access.)

As noted above, we do not know exactly when Tunstall/ Nunawading Station began handling "goods". It was certainly doing so from the late 1940s when Keith Jones was taking boxes of Jones Bros flowers to the Nunawading Station for dispatch to their destination in the city or elsewhere on an electric Parcel Van of Victorian Railways (see Figure 58). This Parcels Van service was available at selected stations from the 1920s and continued well into the 1950s.

Figure 58. An electric Parcels Van of Victorian Railways, circa 1923, in the parcels dock at the western end of No.1 platform Flinders Street Station, Melbourne. These parcels' vans ran around the suburban network, in between normally scheduled suburban electric passenger trains. They were extensively used for moving newspapers and other light parcels, effectively in competition with the P.O. and private couriers/taxi trucks. [Information provided by Phil Rickard]. Photo B41019/163, courtesy of State Library of South Australia.

Clients for the Farm's Produce

The major clients that the farm supplied with corms and bulbs are shown in Table 4. Most were themselves both wholesalers and retailers of corms and bulbs. Similarly, numerous major city and suburban retailers were supplied with cut flowers and foliage (see Table 5).

TABLE 4: SEED MERCHANTS/ NURSERYMEN CLIENTS OF WE JONES AND AJ JONES AND SONS, CIRCA 1910 - 1935

Name of Co.	Full Title	Address of Head Office or Premises
Brown and Co		Station Rd, Malvern, Victoria
Brunn, TH	Brunns Direct Seed Supply Co, or TH Brunn Garden Services Pty Ltd.	39 Queen St, Melbourne, Victoria
Chandler, D		21 Station St, Malvern, Victoria
French, G		241 Queens St, Melbourne
Garford, F		
Hawkins	WR Hawkins and Sons, Plant and Seed Merchants	Box Hill, Victoria
Holloway Bros	Freshwater Nurseries	Manly, NSW
Law Somner	Law Somner, Seed and Plant Merchants	139-141 Swanson St, Melbourne
Montague and Cc	Philip Montague, Bulb Specialist	Humphries Rd, Frankston, Victoria, PO Box 29 Frankston, Tel Frankston 126.
Parker		Clayton, Victoria
Paton and Sons		97 Swanson St, Melbourne
Railton, J	J Railton and Co, Seedsmen and Plant Merchants	271 Swanson St, Melbourne
Watters and Sons	Seed Merchants and Nurserymen	207-09 Swanson St, Melbourne

TABLE 5: RETAIL FLORISTS THAT WERE CLIENTS FOR CUT FLOWERS AND FOLIAGE OF THREE GENERATIONS OF THE JONES FAMILY: CIRCA 1910 - 1959

Name of Co.	Proprietor	Address of Premises
Adele Florists		Royal Arcade
Adrian's		3 Australia Arcade, Melbourne
Astor Florist		536 Burke Rd, Camberwell (Tel Hawthorn 5318)
Bernards Florists		Scotts Hotel, 444 Collins St, Melbourne
Blue Gum Florists	Miss Ramke	Exhibition St, Melbourne (Central 7358)
Buck, E	Miss Buck	Clarendon St, South Melbourne
Cameron	Miss Eunice	127 Exhibition St, Melbourne
Canterbury Bell		Canterbury Rd, Canterbury
Cleary	Miss Cleary	146 Bridge Rd, Richmond
Drew	Miss Drew	Nicholson St, Footscray
Eden Florist		214 A Glenferrie Rd, Glenferrie, (Hawthorn 5104)
Enid Floral Studio		27 Royal Arcade (Central 8823)
Floyd	Miss R Floyd	C Avenue, Eastern Markets, Melbourne
Girrahween Florists		41 Fitzroy St, St Kilda
Hodgins Orchids		
Hotel Australia		Collins St, Melbourne
John Holdsworth		
Jones, E	Miss Elsie Jones	5 Albany Court, 228-232 Collins St, Melbourne, (Central
Kemp	Miss Kemp	Toorak Rd Windsor (or South Yarra)
Lindsays		Flinders Way (Central 2364)
Lucas	Miss Olive Lucas	7 Block Court, Melbourne
Margo Ann	Mrs. J Harwood	Exhibition St, Melbourne
Millefleur		The Block, 300 Collins St, Melbourne
Noel's Florists		Little Collins Street, Melbourne
Oak Leaf	E Petersen	Oakleigh
Paton and Sons		97 Swanson St, Melbourne
Ronalds Florists		229 Collins St, Melbourne
Royle Florist		Windsor
Saks Flowers		Myer Emporium, Bourke St, Melbourne
Sharp	Miss Sharp	
Sipthorpe, A		High St, Armidale
Suzanne Gay	Mrs Cook	Whitehorse Rd, Mont Albert
Van Breda		91 Fitzroy St, St Kilda
Wood, M	Miss Maisie Wood	Centre Court, Melbourne
Wright, AC		302 Smith St, Collingwood

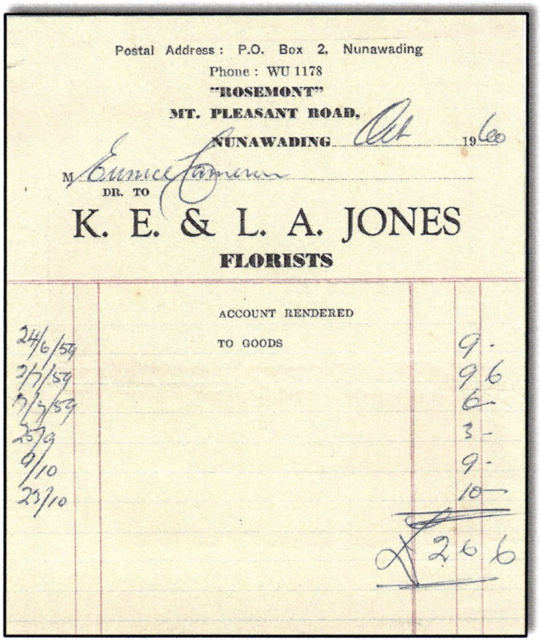

Postal Address: P.O. Box 2, Nunawading
Phone: WU 1178
"ROSEMONT"
MT. PLEASANT ROAD,
NUNAWADING Oct 1960

M Eunice Cameron
DR. TO

K. E. & L. A. JONES
FLORISTS

ACCOUNT RENDERED
TO GOODS

24/6/59		9	·
2/7/59		9	6
17/7/59		6	–
25/9		3	–
9/10		9	·
23/10		10	–
		£2	6 6

Figure 59. Possibly the last invoice sent by KE and LA Jones to a loyal retail customer (Eunice Cameron) for miscellaneous flowers supplied during 1960. The handwriting is Keith's.

Figure 60. Rosemant Farm from the bedroom window, late 1950s.

CHANGING TIMES

We regard the Jones flower growing/nursery story as one of impressive innovation and adaptation to changing circumstances (or perhaps responsiveness to emerging opportunities) by four families of Joneses.

The Walter Jones Years

In the 1890s, Walter Jones realised that his business plans would require more land than was available in and around his father's property in Preston so he acquired 20 acres of land at Tally Ho/ East Burwood. Here, we believe, he saw an opportunity to supply seedling trees to the rapidly expanding apple and pear orcharding businesses in the locality, to capitalise on an export opportunity for these and other fruits in South Africa, and to grow some flowers.

Then, sensing that that there were better opportunities in the Melbourne bulb and cut flower business, particularly if he had access to a good public transport, he sold his East Burwood property and moved to 14 acres of land at Tunstall close to the railway line. Fortunately, he came across a property that had been partially developed 20 years before by someone who described himself as a vigneron.

TABLE 6: FREQUENCY OF THE WORDS "DAFFODIL" AND "ORCHID" IN PRINCIPAL NEWSPAPERS (THE AGE AND THE ARGUS) IN THE PERIOD FROM 1890 TO 1960**

Decade	**Number of "Hits"**	
	Daffodil	Orchid
1890-1899	676	1,459
1900-1909	1,269	994
1910-1919	1,484	1,052
1920-1929	1,619	1,556
1930-1939	**3,044**	3,811
1940-1949	1,156	**4,125**
1950-1959	908	3,362

***The search used the digitised newspaper section of the website www.nla.gov.au/trove. The peak number of hits for each species is shown in bold.*

Over the next 15 years he further developed the infrastructure of this property, adding several new buildings and developing the capacity to irrigate plants during dry times via a system of dams, windmills and water tanks. The large purpose-built Bulb Shed that we believe he had built suggests that the production and wholesaling of the bulbs of daffodil and related Narcissus species was a very important part of his business. In effect, he predicted the rise in the popularity of this genus, which grew, steadily to its peak in the 1930s (see Table 6).

However, cut flowers must also have been important to Walter's business, given the extensive plantings of Roses, Camellias, Prunus, Japonica, Hydrangea and Bugle Lilies (Watsonia) that we believe he made during his tenure.

Tragedy and advancing years, however, brought the Walter Jones years to a close. The death of two of his sons in the Great War put an end to any plans he might have had of passing the property on to them, so he sold it as a going concern to his youngest brother, Arthur Jones, in 1917.

The Arthur Jones Years

In 1917, 44-year old Arthur probably had little experience in the flower and nursery business, so it was perhaps fortunate for him to have an already established business to walk into. The evidence suggests that he just continued running the existing business and did not further develop the infrastructure to any great extent. He continued with daffodil bulb growing and wholesaling on quite a large scale, employing considerable casual labour in support (see Table 3). Examination of Table 6 suggests that he was in the "right business at the right time" given the steadily increasing popularity of daffodils and related Narcissus species and cultivars that reached its peak in the 1930s.

Bulb production, however, formed only one part of Arthur's business as he and his two sons also supplied cut flowers (including daffodil blooms) and foliage to the Melbourne retail florist trade during this period (see Table 5). He probably also planted quite a few more flowering trees and shrubs, particularly roses. In fact, the two rose cultivars that were the mainstay of the farm's cut flower business right through to the late 1950s - "Black Boy" and "Lorraine Lea" - were bred at Bulla, Melbourne by Alister Clarke in 1919 and 1924 respectively.

Arthur retired in his early 60s in about 1935, transferring then selling the business to his sons Keith and Lionel. He then moved to Black Rock on Port Phillip Bay, returning to the farm during 1942 to manage it while Keith and Lionel were away during the war.

The Keith and Lionel Jones Years

1935 TO 1940

Keith and Lionel were clearly in charge of the farm from 1936. In the first three or four years (i.e. until Lionel volunteered in 1940) they presumably continued earning their income by supplying cut flowers and foliage to their father's retail customers, while beginning the process of dramatically changing the nature of the business - phasing out wholesale bulb production altogether, investing in glasshouses for an expanding collection of orchids, and planting more roses. Phasing out bulb production may have even commenced in the early 1930s, i.e. towards the end of their father's tenure, since the farm records for labour terminate in about 1932 and for bulb sales in about 1935.

Keith and Lionel successfully foresaw the growth in popularity of orchids that peaked in the 1940s (see Table 6) by steadily expanding their involvement from a hobby interest in the late 1920s to a major component of their business by the 1950s. It should be noted that orchids were relatively rare in those years to such an extent that, if a woman received a corsage containing orchids from her beau, she considered herself to be highly flattered and very special!

1940 TO 1946

The war years put the business under great stress and brought many changes. Lionel joined the army in 1940 and went off to the Middle East; Keith continued on as best he could with some help from his father on his weekly visits. Then he also joined up and spent much of the next two and a half years away with the RAAF. During this time, Arthur and daughter-in-law Elma did their best to keep the business going. Keith was discharged on compassionate grounds in September 1944 because of his father's declining capability, and in 1946 Lionel returned (from the war in Libya, imprisonment in Italy, and internment in Switzerland) and he and Keith set about rebuilding the business.

1946 TO 1959

This period following Lionel's return to Australia was one of great activity on the farm as orchids and roses became much more central components of the business. Keith and Lionel had several more glasshouses built for their expanding production of slipper, Cattleya and Cymbidium orchids. They also targeted the winter demand for roses, particularly the deep red "Black Boy" cultivar, and developed special techniques for satisfying this demand from the three of four long rows of this cultivar. This involved a hard pruning in early autumn (March- April), rather than the normal winter pruning in July/ August. This gave the bushes some time to make new growth and produce flower buds before the cold weather arrived. As a consequence, the bushes flowered in late winter and early Spring – a few weeks earlier than winter-pruned bushes. Rows were then well fertilised with chicken manure. Other rose cultivars such as Lorraine Lea and Mlle Cecile Brunner continued to produce their usual flower crops, even though many of the plants were 20-40 years old. When the lovely rose cultivar, Peace, became available in the late 1940s, they planted a row of standard plants (i.e. with a metre-long erect stem) since this made picking and weed control easier.

[The history of this cultivar is particularly interesting. Peace, more correctly named 'Madame A. Meilland', a well-known and successful hybrid tea rose, was developed by French horticulturist Francis Meilland in the years 1935 to 1939. When Meilland foresaw the German invasion of France, he sent cuttings to friends in Italy, Turkey, Germany, and the US to protect the new rose. It is said that it was sent to the US on the last plane available before the German invasion. Here it was safely propagated by the Conard Pyle & Co. When the war ended, it was sent back to Europe and given a variety of names in the various countries. The cultivar has large flowers of a light yellow to cream color, slightly flushed at the petal edges with crimson-pink. It is hardy and vigorous and relatively resistant to disease. Information from Wikipedia.]

However, Keith and Lionel also continued supplying retail florists in Melbourne and suburbs with quite a wide range of cut flowers and foliage including prunus blossom, japonica, hydrangeas, watsonia, lavender, azaleas, camellias, and rhododendrons as well as the flowers of numerous other more minor plants. While, they were able to cut flowers and foliage from these old plantings of trees and shrubs, they also planted extensive areas of annuals every year, particularly stocks (Matthiola incana) and gladioli for the late winter/spring flower trade and dahlias for the summer trade. The fact that they were able to cut their flowers in the early morning, pack them in boxes and rail them to retail outlets before mid-day, gave them the key advantage of freshness over bulk-produced flowers sold to retailers through the Queen Victoria Markets.

Keith and Lionel followed a fairly standard weekly work routine. One of us (Roger), the middle child of Keith and Elma, spent his first 22 years living on the farm, and has written elsewhere about what he remembers about the weekly activities on the farm in the 1946-59 period, as follows:

"On a weekday, Keith and Lionel got up at about 5 AM when they had heard the first train and started picking the flowers for the day's orders. These first few hours of the day were pretty hectic as they progressively filled the boxes for the various customers. I think they would also pick whatever was in season and include it because a good florist could always find a use for nice flowers and foliage that had actually been picked that day. Then it was a quick breakfast before harnessing our draught horse "Nugget" and loading up the cart with the boxes, and heading for the Tunstall/Nunawading

station. Here Keith waited for the railway Parcel Van and helped load the "Jones Bros, Tunstall" boxes onto it. After that, he did the shopping at the store, grocery, and butcher on the northern side of the railway line. On the way home, he would sit up on one corner of the cart to read the Age newspaper and let "Nugget" have his head; he, of course, obliged by getting slower and slower so the one-mile journey from the station could take 30 or 40 minutes. This trip to the station took about two hours in all – during which time Lionel did odd repair jobs.

After about 10:30 AM, the actual farm work would commence – pruning roses, hoeing weeds, mowing, ploughing land for a new crop (Keith was always on the plough handles and Lionel led the horse and there was much swearing and cursing), applying manure, working in and around the glasshouses, planting annuals, shifting hoses, moving "Nugget" or the cow - both of whom were tethered, collecting firewood from the bush area, or whatever the season dictated. As on any farm, there were some busy times and some pretty slack times when nothing much got done. No work was done on a Saturday afternoon or Sunday. Sunday afternoon was a time for visitors to walk around the farm, look at the orchids, and have afternoon tea. There were many Sunday visits from other orchid growers and enthusiasts."

Keith and Lionel seemed reluctant to invest funds in the maintenance of the farm's infrastructure, particularly its buildings. Keith and Elma gave financial priority to the education of their children at top Melbourne private schools whereas Lionel, being unmarried for much of this time, invested his money in a block of shops in Watsonia. Thus, they spent very little money on maintenance of the buildings (e.g. the Stables/ Feedshed, the Packing and Bulb Sheds, and the shade structures for Hydrangeas) with the result that they continued to deteriorate. Similarly, they did not modernise their land preparation methods by investing in say a small tractor e.g. a Ferguson or "Fergie" or a new Rotary Hoe, preferring to rely as they had since the 1930s on the single mouldboard plough and a set of harrows pulled by "Nugget" and a rather worn out Howard rotary hoe. Their emphasis was on producing and selling whatever flowers and foliage the farm could grow without much expenditure.

It was as if they knew that flower growing on the farm was coming to an end.

This was in fact what happened. By the late 1950s, the rapid post-war expansion of Melbourne suburbs began to overwhelm the flower and fruit growing areas of Nunawading, and Keith and Lionel realised that the pressures from rising council rates and demands from developers were irresistible. In their case, it was the Education Department that made the first move. In about 1957, the Department compulsorily acquired most of the "bottom garden" of the farm for what was to become the Nunawading South State School. Soon after this transaction was complete they began negotiations with a developer over the sale of the remainder of the farm. These were completed by about 1959 so they sold off as much of the infrastructure as they could (plants and some structures such as parts of glasshouses, the oil burner, etc.) and vacated the house and land in 1960.

The family's connection with the house did not end there, however. Keith's daughter, Helen, and her new husband, Allen Sheridan, decided to move it to Bairnsdale in East Gippsland. Allen spent many hours removing the plaster from the lathe and plaster interior walls and demolishing the chimneys in preparing the house for removal. The old house was deposited at 119 Moroney St, Bairnsdale and after much additional work became their first family home where they lived happily for the next several years. The now ~117-year-old house was still occupied in 2019.

Roger has clear memories of this end of the Jones flower-growing era.

> *"In 1960, someone in the family had organised for Arthur's three older sisters to visit him and his second wife Glad at their home at 65 Mt Pleasant Rd and to have a final look at the flowers on the farm. At the time, Arthur was about 87 years old, Mauder 89, Avonia 93, and Edith 98. My job was to take the photographs (see Figure 62) of this historic occasion, as I was the only one in the family with an interest in photography. I can remember thinking that I was seeing, probably for the last time, the last survivors of the 10 children that James had fathered between about 1840 and 1873. It was close to the end of that generation of Joneses.*
>
> *Not long after this occasion, it was time for Mum and Dad (Elma and Keith) to move out of the old house. By then, the farm was in disarray as many of the plants that could be moved and some of the infrastructure had been sold and taken away. The farm was only a shadow of its former glory. On the morning of their departure Dad and Mum stood at the back of the furniture truck and watched the last of the furniture loaded. Dad was silent; he had spent 43 years of his life on this block of land and knew every inch of it like the back of his hand and every plant that had grown on it. Mum was in tears as she silently recalled the 26 years she had spent there. She had come as a young bride (Figure 61), had endured the hardships of the war years here, had actively participated in the business, had raised three children here, and had cooked, sewed, darned and laundered for them and for herself and two men – her husband and her brother-in-law.*
>
> *It was truly also the end of the era of Jones family members participating in commercial plant and flower growing".*

Miscellaneous photos of Rhododendron, Dahlia and Camellia flowers.

Figure 61. Keith and Elma were married at Kooweerup Presbyterian Church in December 1935. This photograph shows them with Lionel and Elma's sister, Dulcie Jeffery.

Figure 62. Arthur and his three sisters Mauder, Avonia and Edith, cira 1960.

Fortunately, both Keith and Lionel were near retirement age so they each went their separate ways. Lionel purchased a Lorikeet Street block from the developers in 1960, near the old farmhouse, to keep the family connection. His own house was on two adjacent blocks that he had purchased in the late 1950s from TM Burke (the estate agents for the old Moore orchard land). Elizabeth was born in 1960 and she and her husband Murray Ogden eventually built a house on this Lorikeet Street block. They lived there for 20 years, selling it in 2013.

Figure 63. (Right) A poster for the 1960 Spring Show of the Victorian Orchid Society. The orchids illustrated are Cymbidiums.

Figure 64. (Left) Lionel, as President of the Victorian Orchid Society in 1970, with the Grand Champion of the Spring Show awarded for a Dendrobium exhibited by R Hodgins.

In retirement, Lionel developed the gardens on the two blocks of land on which his new house stood, and re-erected part of a glasshouse from the farm growing the orchids that he really liked. He also became even more active in the two Orchid Societies as both a judge and office bearer. He was President of both on a number of occasions.

Keith and Elma retired initially to a hillside block with a spectacular garden at Kalorama in the Dandenong Hills and after a few years moved to a small farm at Kalimna West near Lakes Entrance in East Gippsland to be near their daughter Helen.

The retirement of Keith and Lionel marked the end of a 70-year involvement in flower growing of three separate generations or families of the Jones family.

Figure 65. Keith and Elma's three children, Jeffery, Roger and Helen

Figure 66. Lionel and Bette with baby Liz.

Although the Jones family's commercial involvement with flowers ceased in 1960 with Keith and Lionel's retirement, the love of flowers, plants and gardens continues on through their children to this day, almost 60 years later.

- Jeffery has in recent years been busy beautifying the streets of Kalgoorlie-Boulder in WA with spectacular pots of geraniums; his publication about this work can be found and downloaded at: http://geraniumcottage.com/doc/Cascading_Geraniums_Booklet.pdf.
- Roger continues to convene the gardening group at his local Church and helps maintain the gardens and kerbside at his apartment complex in Brisbane;
- Helen runs a large farm garden and orchard at their farm near Bairnsdale, Vic; and
- Liz has recently expanded her shadehouse for orchids from her father's collection at their home in Blackburn, Vic.

ACKNOWLEDGEMENTS

This book started out as a very factual description of the land that the four Jones families purchased after about 1890, the plants they grew, the flower farms they developed, and the business models they followed. A close friend, Dr Martin Playne, got us off to an excellent start in researching some of the land transactions and that added greatly to the story. His work was continued and expanded on by Liz and her friend, Anne Jones, (no relation)

Then another close friend, Dana McCown, offered to assist us in humanising a story that read like a scientific report, and moving it towards a well-rounded publication. She pressed us repeatedly to tell the parallel stories of the four families involved and was not happy until we had found the photographs and life stories of many of them. We are extremely grateful to all three, Martin, Anne and Dana.

We are also very grateful to our siblings or cousins, Helen Sheridan, Jeffery Jones, and John Coghlan for their recollections of plants and activities on the farm in the 1940s and 50s and for the access they gave us to the wider family's collections of photographs.

In fleshing out the people stories, we have been assisted greatly by access to the digitised newspapers that are available from the National Library through their website *www.nla.gov.au/trove*. We are also grateful for access to records held by the Public Records Office of Victoria (PROV), The State Library of Victoria (SLV) and Victoria's Department of Environment, Land, Water & Planning (DELWP) and, of course, to the wonderful *Wikipedia.*

APPENDIX ONE

The subdivision of Crown Allotment 116 from 1864 to 1905

Readers may be interested in the progressive subdivision over time of the original 331 acres of Allotment 116.

Our starting point is the 1864 map of the Nunawading Parish that showed that much of the land bounded by what was to become Whitehorse Rd, Mitcham Rd, Canterbury Rd, and Springvale Rd was Crown Land Reserve in three Allotments: Numbers 116, 117, and 118. These were of about 331, 112, and 115 acres respectively (see Figure 2 in the main document).

By 1878 (see Figure A 1), Allotment 116 had been divided into six smaller holdings in the 20 to 80-acre range and owned by J Wakefield, T Jobbins, FA Volgt, D Boyle, P Geraghty, and A Rooks. 116 G and 116 D contained the two further subdivisions which constituted the final Jones Flower Farm.

(Part of the Wesleyan Church block is still occupied by the Wesleyan Church which became the Uniting Church of Australia in the 1970s. Keith and Elma's children attended Sunday school and Church Services at this Church in the 1940s and 1950s. In about 1948, Roger and Helen also attended grade 3 and 4 school classes in the Church Hall because the student capacity of the Forest Hill State School had been exceeded. Helen was married in the Church and had her wedding reception in the Hall in 1960.)

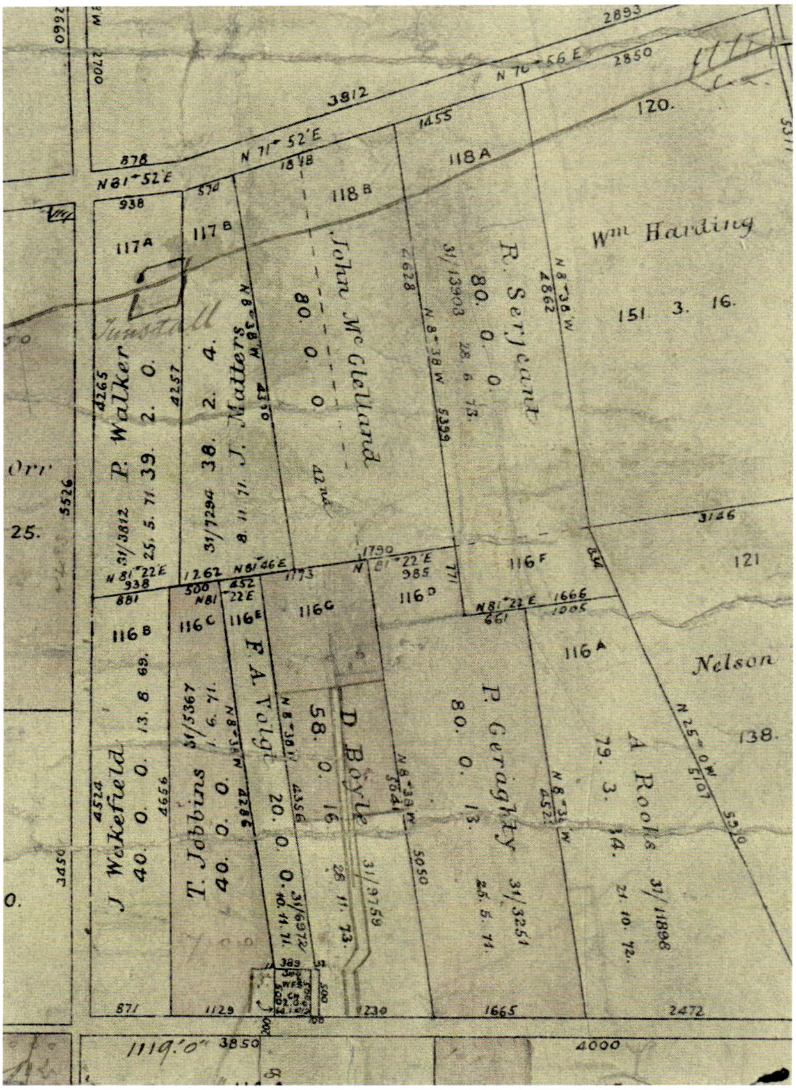

Figure A 1. A portion of the "Map of part of the Parish of Nunawading, 1878, photo-lithographed at the Department of Lands and Survey, Melbourne by TF MacGauran". This part of the map shows the lands bounded on the west, south and north by the future Springvale, Canterbury, and Whitehorse Roads.

Note that the "Boyle block" is 116 G despite the "G" looking much like a "C" on this map. Note also that the southern portion of Allotment 116 E, owned by FA Volgt, was held by the Wesleyan Church. http://handle.slv.vic.gov.au/10381/167865 and http://search.slv.vic.gov.au/MAIN:Everything:SLV_VOYAGER2178759).

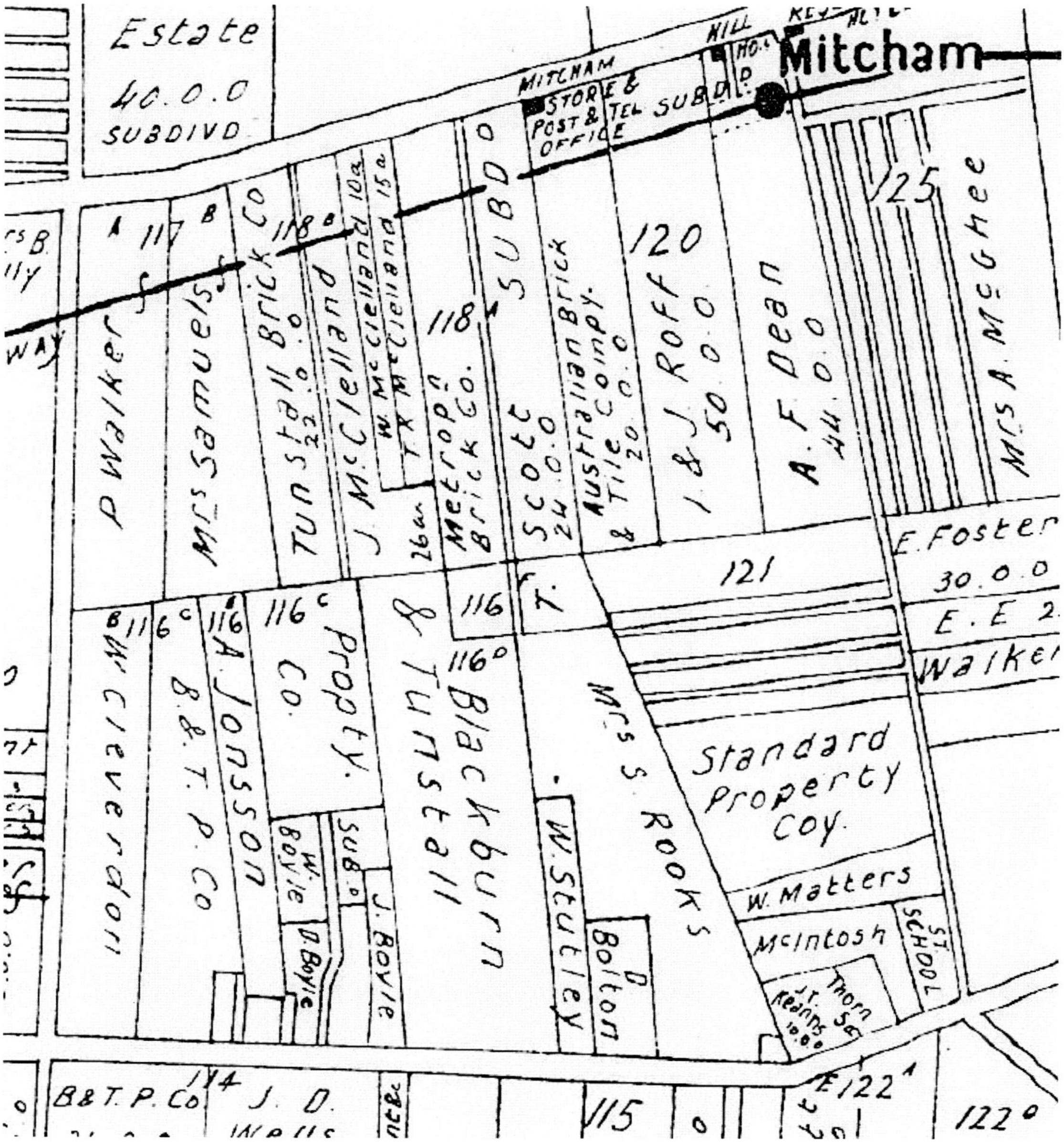

Figure A 2. The land ownership situation in 1892 as shown in this portion of a map in the Municipal Directory of 1892 reproduced in Brennan's "The history of Nunawading" (1972). The northern section of 116 G (looks like a C on the map above) contained the future "bottom garden". The northern section of 116 D contains the future "top garden" of the Jones flower farm.

ALLOTMENT 116 G

David Boyle of Box Hill had advertised in the South Bourke Standard (Vic.: 1861 – 1873) on Friday 24th October 1868, page 2 that "he had applied, under the 42nd Section of Amending Land Act 1865, for licenses to occupy three twenty acre blocks of land situated in Allotment 116, Parish Of Nunawading". This request was granted and he became the occupier of Allotment 116 G consisting of approximately 61 acres in 1868. He subsequently purchased his selection from the Crown on the 28th of November 1873 for £59. Note that the southern part of 116 G (see Figure A 1) had a central roadway running approximately N-S and several potential sub-divisions pencilled in, suggesting his intention to subdivide. Boyle cleared some of the land and built a wattle and daub house there on what became known as Scotsman's Hill, later Forest Hill, and established his garden and nursery.

Boyle sold the northern half (No 161218) on the 25th November 1885 to the Freehold Investment and Banking Company of Australia Limited and subdivided the balance on the 21st August 1886 to various family members – probably grandchildren - (Walter Boyle, William James Boyle, and Joseph Nash Boyle), creating an easement north from Canterbury Rd in the process (the future Mt Pleasant Rd). Other blocks were sold in March 1894 to Janet Kennedy Alexander, Joseph Nash Boyle, and Walter Boyle.

The northern half of 116 G, shown as 161218 in Figure A 3 was further subdivided between 1900 and 1906 (see Figure A 4). In the process, the easement that was to become Mt Pleasant Rd was extended even further to the north. Similarly, an E-W easement (originally called Thiele's Lane but now an extension of Eugenia St) was created.

One block from this second subdivision of 116 G (486911 – see Figure A 4) was purchased on 14 April 1904 by WE Jones. It became the "bottom garden". Other blocks were purchased by Charles Course (11 December 1900); Alfred John Williamson (2 February 1905); Arthur Henry Plumridge (23 February 1905); Joseph Boyle (2 October 1905); and Ambrose Fredrick Thiele (11 April 1906).

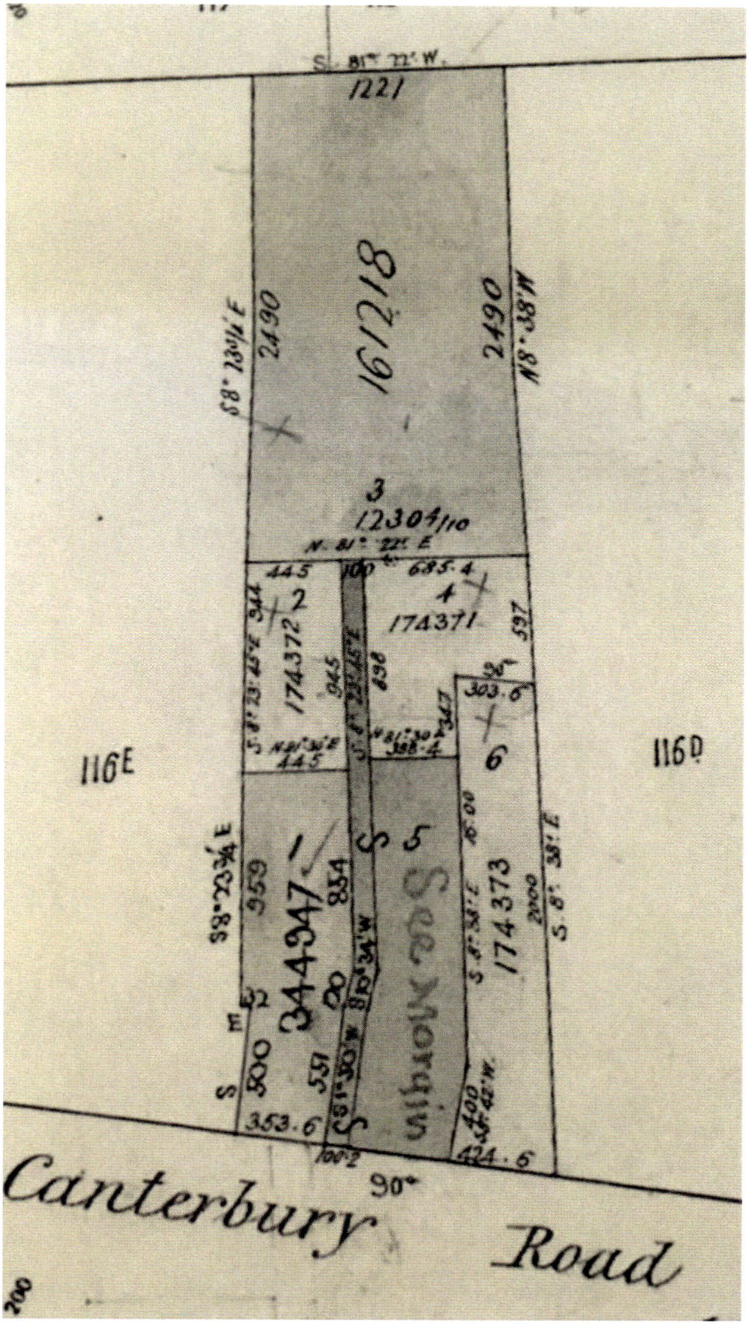

Figure A 3. The first formal subdivisions of David Boyle's Allotment 116 G, dated 1886, and the development of an easement north from Canterbury Rd that was to become the southern section of Boyles Rd, later Mt Pleasant Rd. The block that was later purchased by Walter Jones to become the "bottom garden" was within the top or northern block (No 161218). Information courtesy of Victoria's Dept of Environment, Land, Water & Planning (DELWP) via the Public Records Office (PROV).

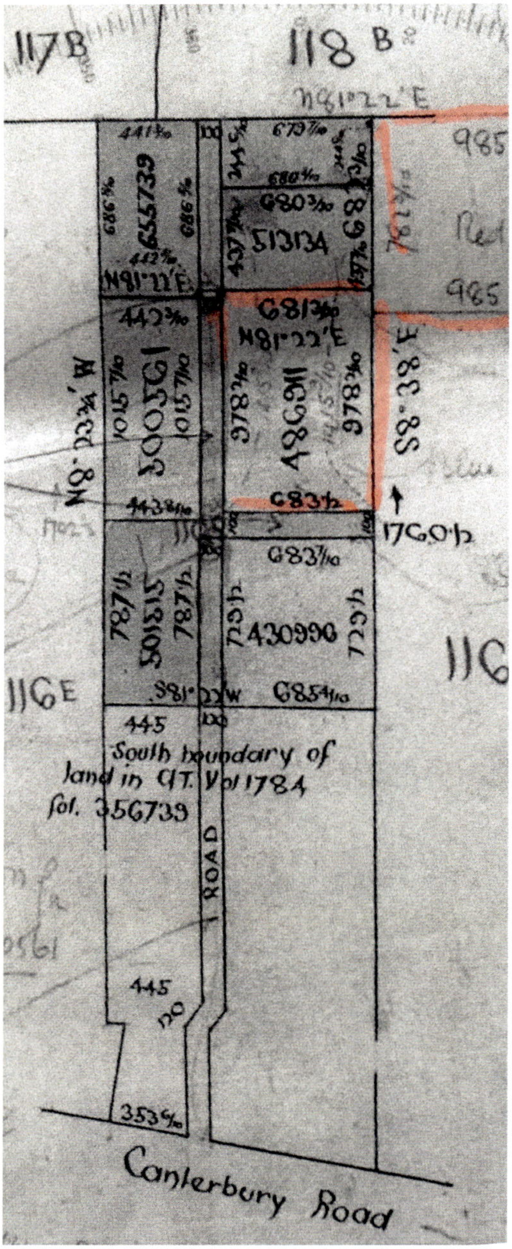

Figure A 4. The second major subdivision of the northern block of Allotment 116 G between 1900 and 1906. Block 486911 (outlined in pink) was purchased by Walter Jones and became the "bottom garden" of his flower farm. Measurements on the sides lengths in links (1 link = 20.1 cm). Information courtesy of Victoria's Dept of Environment, Land, Water & Planning (DELWP) via the Public Records Office (PROV).

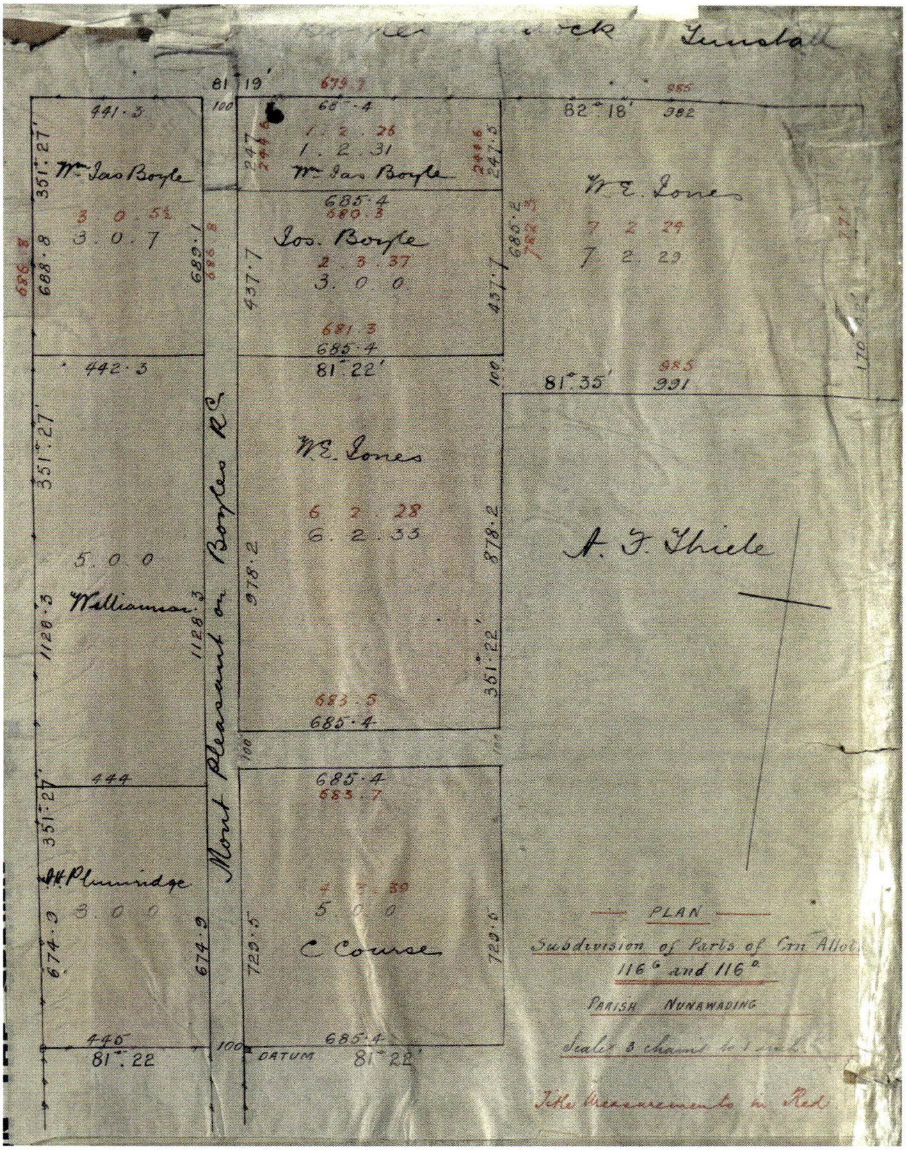

Figure A 5. Hand-drawn survey "Plan of the subdivision of parts of Crown Allotments 116 G and 116 D, Parish of Nunawading" circa 1900, held by the State Library of Victoria. The heading on the top of the page appears to say "Boyles Paddock, Tunstall".

REFERENCES: http://handle.SLV.VIC.GOV.AU/10381/169580

(Length measurements are in chains and links, compass bearings are in degrees and minutes, and areas are in acres, roods and perches.)

ALLOTMENT 116 D

This part of the Allotment (see Figures A 1 and A 2) contained the most easterly of the two blocks that made up the final Jones flower farm. It was originally purchased from the Crown on the 25th of May 1871 by Patrick Geraghty of Nunawading for £81. This approximately 83-acre allotment was bounded on the south by the future Canterbury Road. (A small rectangular portion on the north-east corner of the roughly rectangular Allotment 116 D had been previously excised as part of Allotment 116 F.) Patrick Geraghty mortgaged Allotment 116 D several times in the early 1870s to the Land Mortgage Bank of Victoria Ltd, then transferred it to William Geraghty, presumably a family member, on the 31st of May 1877.

On the 14th of August 1877, however, Peter Boldini, a vigneron purchased Allotment 116 D. A search of the Municipal Rate Books for the period 1879 to 1886 confirms that Peter Boldini, a farmer, was the owner from 1879 to 1884. His neighbours on the future Mt Pleasant Road were David Boyle (farmer, nurseryman), Joseph Boyle (labourer), William Boyle (labourer, later contractor), Walter Boyle (chair maker), Charles Course (described in 1905 as a florist), and Arthur Henry Plumridge (also described in 1905 as a florist).

(The possible significance of Boldini's ownership of this land on the farm's infrastructure is discussed in the main document.)

Boldini soon seems to have got into financial trouble that required him, between 1877 and 1882, to mortgage his 83 acres several times to various individuals and finally to the Australian Deposit and Mortgage Bank Ltd. Then, on 10 February 1883, Boldini advertised his farm in the Melbourne Age for sale by Auction on 24th February (see Figure A 6). The auction was apparently unsuccessful but the land was finally sold or transferred on 4th August 1885 to the Freehold Investment and Banking Company Australia Limited. Clearly, these were very turbulent times financially!

This company held the land until 14 April 1904 when, in receivership, it sold a small northern section to WE Jones and, two years later, the balance to Ambrose Fredrick Thiele (see Figure A 5) and it was in his ownership until his estate was wound up in 1951. This Thiele land shown as 523273 in Figure A 7 was to become the Moore Orchard that grew apples and pears until the mid-1950s when it was again extensively subdivided into the current suburban residential blocks.

[Interestingly, the 1892 map (Figures A 3) showed Allotments 116 D, 116 C and the northern part of 116 G belonging to the Blackburn and Tunstall Property Company, but the title documents don't support this ownership change – perhaps they held the land between the auction and August 1885? According to Fig A3, other Allotment 116 block owners were W Cleveland; A Jonsson; W, D, and J Boyle; W Stutley; D Bolton; and Mrs S Rooks.]

SATURDAY, 24th FEBRUARY.
At Half-past 4 0'Clock.
On the Property.
MOUNT PLEASANT,
12 Miles from Melbourne, and 1 ¾ Miles from Mitcham
Railway Station,
On the Lilydale Line,
CANTERBURY-ROAD,
Close to Church, Post Office and State School,
80 ACRES LAND,On the Crown of the Hill, View of the Bay.
WEATHERBOARD COTTAGE,
Four Rooms, Cemented Tank, Outhouses, Stabling.
Title Freehold.
BEAUCHAMP BROTHERS have received instruct-
tions from Mr. Peter Boldini to sell by auction,
on the property, on Saturday, 24th, at half-past 4o'clock,
His very comfortable
FARM,situated one mile from the Mitcham station,
by the paddocks, or 1 ¾ miles by the roads.
The position is charming, commanding a
grand view of the bay, being on the crown of
a hill. The locality is well known, and recom-
mended for its salubrity.
80 ACRES LAND,
all well fenced with posts and rails, live
hedge in front, three paddocks fenced with
acacia hedge, fairly timbered, 25 acres under
cultivation, three acres orchard, containing
some of the best fruit trees in the colony.
SUBSTANTIAL WEATHERBOARD COTTAGE,
containing four good-sized rooms, also out-
houses, stabling, large cemented tank.
Good fishing and shooting close to the property.
A never-failing creek runs through portion of the land.
Stock and working plant can be taken at a valuation.
Train leaves Melbourne at 20 minutes to 4 o'clock,
and arrives in time for the sale.
Plan and fuller particulars of the property obtain-
able from the auctioneers.
Title, Crown grant.

Figure A 6. Advertisement in the Melbourne Age for the 80-acre property at Mt Pleasant owned by Peter Boldini. (Source: The Age (Melbourne, Vic 1854-1954), Saturday 10 February 1883, Page 2.)

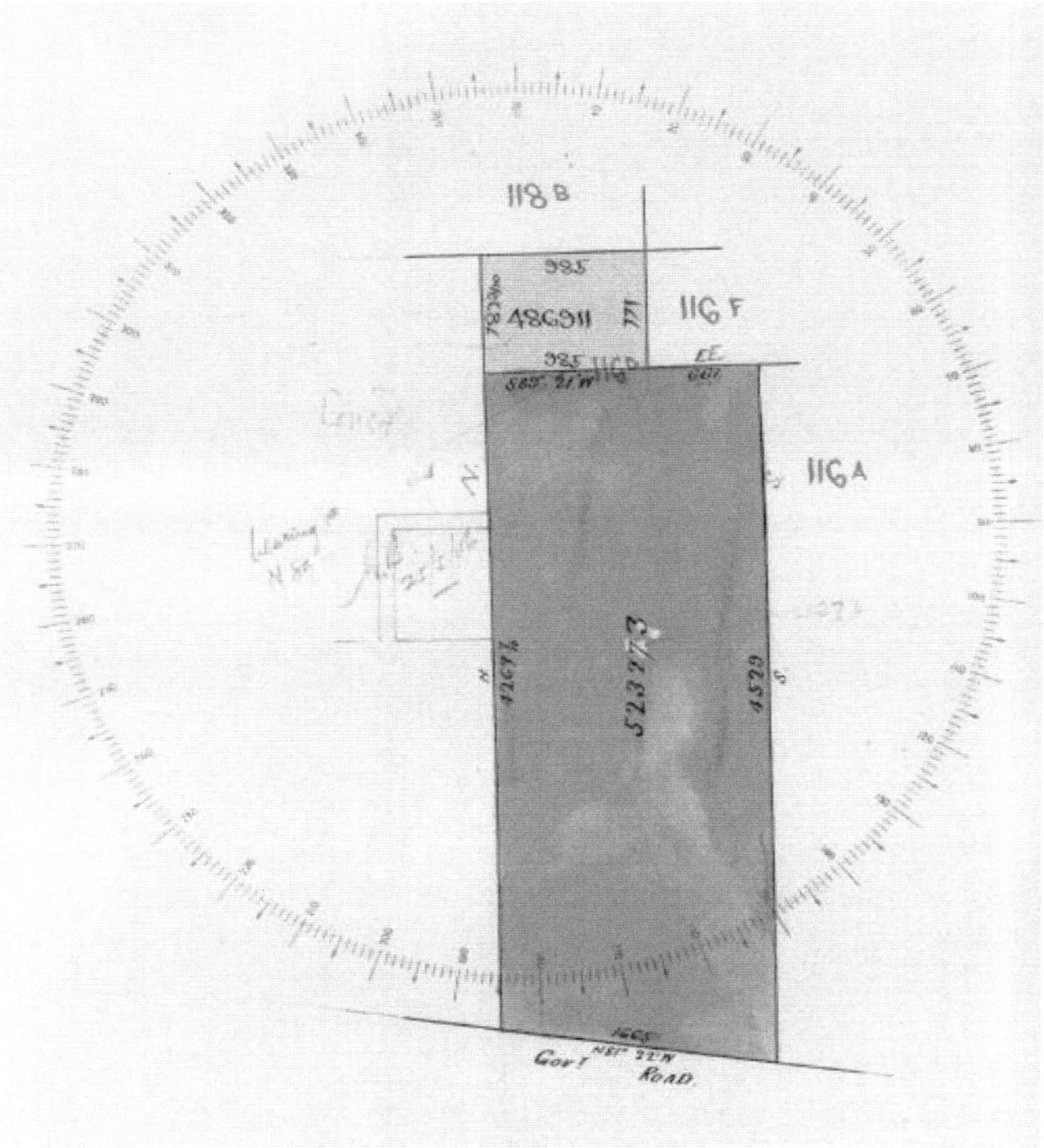

Figure A 7. The second subdivision of Allotment 116 D. The northern section (486911, 7.65 acres) was purchased in 1904 by Walter Jones and became the "top garden". The southern section was purchased in 1906 by AF Thiele. In time this became Moore's orchard. Information courtesy of Victoria's Dept of Environment, Land, Water & Planning (DELWP) via the Public Records Office (PROV).

The two portions of Allotments 116 G and 116 D that Walter Jones purchased in 1904 were amalgamated into the one block as shown in Figure A8.

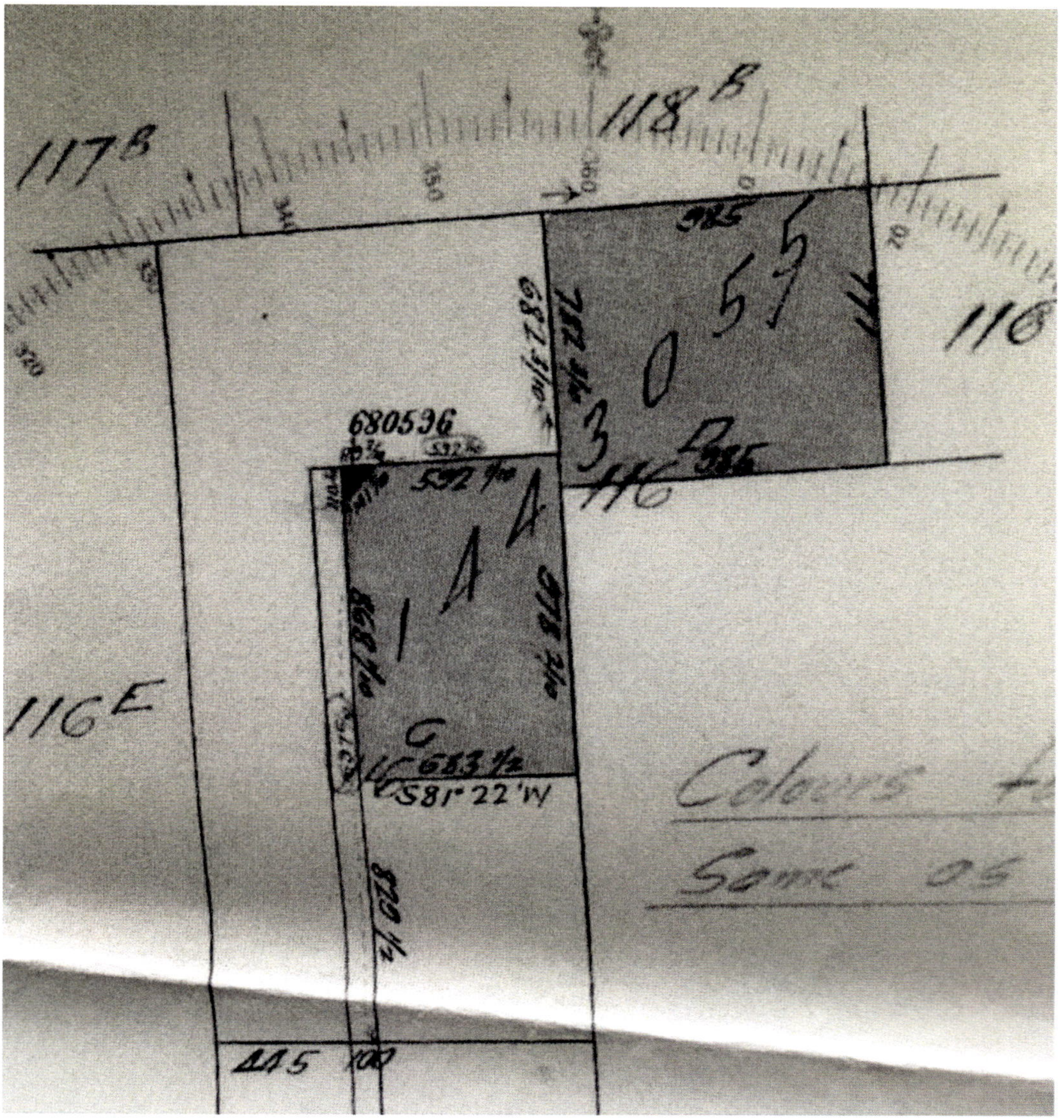

Figure A 8. An excerpt from the 1904 title documents (Vol 2983, Fol 596471) for the two blocks of land purchased by Walter Jones and amalgamated to become the original "Rosemont" flower farm. The most westerly block fronts the future Mt Pleasant Rd that was developing from the south. Note that a) measurements on sides of blocks are in links; b) the two blocks butt against each other for a very short distance (100 links or 20.1 metres); and c) the bottom block (the "bottom garden") facing the future Mt Pleasant Rd had a small triangular section in its NW corner excised for what was, or was to become, the "MMBW's "Pipe Track" for transferring water from the Mitcham reservoir to Melbourne suburbs.

Information courtesy of Victoria's Dept of Environment, Land, Water & Planning (DELWP) via the Public Records Office (PROV).

APPENDIX TWO

David Boyle, Prominent Early Tunstall Resident

David Boyle (1821-1900) was a Scot who came to Australia in 1841. He had received formal botanical training and was associated with the Scottish Horticultural Society. A gardener and nurseryman and a keen collector of new Victorian plants that he came across in his travels, he often did sketches and paintings of their flowers and fruits. As detailed in Appendix 1, Boyle purchased Allotment 116 G of 61 acres in the Shire of Nunawading for £59 in November 1873, having selected it five years earlier. He cleared some of the land and built a wattle and daub house there on what became known as Scotsman's Hill, later Forest Hill, and established his garden and nursery.

In the 1860s, Boyle became particularly interested in the plants of the nearby Dandenong ranges, particularly the tall eucalypts. An article in the Australasian (Melbourne, Vic: 1864 – 1946), on Saturday, 24 November 1866, page 8 refers to his work under the heading:

MISCELLANEOUS NOTES AND QUERIES – Measurements of Australian trees –

> We have received the following memorandum from Dr Mueller*: -
>
> *"Whether the coniferous trees of California or the myrtaceous trees of Australia rank as the most colossal of the globe is yet a question open for discussion. Lately a tree of Eucalyptus amagdalina (the peppermint, and also the drooping gum-tree of Tasmania) was measured in the deep recesses of Dandenong by Mr. D Boyle, of Nunawading, who found the length from the base of the stem to where the upper part of the branches had broken to be 392 ft. Comparing trees uninjured, he assumes that about 30 feet have to be added as a fair estimate of the length of the tree before it dropped, giving us the stupendous height of about 420 feet for this colossal species, while 450 feet is the greatest an exceptional height on record for the Californian Wellingtonia. ....*
>
> *It would be interesting if measurements were made by residents in various parts of Australia of the particular kinds of trees of their neighbourhood, of whatever kind such trees may be, in order that they may be recorded before the trees themselves sink under the acts of the woodcutter, and finally become annihilated."*

(*Baron Ferdinand von Mueller was the first Director of Victoria's Royal Botanic Gardens)

This article started a controversy, which raged for the next 30 or more years and particularly around the time of the 1888-89 Melbourne Centennial International Exhibition that was aimed at showcasing Victoria's achievements to the world. In effect, Victoria was challenging California's claim to have the tallest trees in the world so the State's honour and reputation were at stake! Boyle was not to be silenced and in 1888 he made further claims. Thus:

The Herald (Melbourne, Vic.: 1861 – 1954), Friday, 31 August 1888, page 3 under "Exhibition Notes" states as follows

> *"An extraordinary communication has been received by the Commissioners from Mr. David Boyle, Forest Hill, Nunawading, in connection with the discussion as to the representation of big Victorian trees at the exhibition. Mr. Boyle claims to have discovered, in his neighbourhood, the largest tree in the colony, and his letter is accompanied by a remarkable sketch of the tree in question, with ample notes. .... The height of the tree, Mr. Boyle states, is 524 feet, and its description is supplied as follows "Nat Ord, Myrtaceae, Eucalyptus, Amddalina. ... The circumference of the trunk 10 feet from the ground is stated to be 73 feet. If this monster of the forest is as large as he is represented to be in the description, Mr. Boyle has certainly made a discovery." And so the claims and counter-claims continued."*

With the passage of time and more rigorous measurement, the controversy was resolved. Although some of the Victorian Eucalypts were indeed massive trees 300 or more feet high, Boyle's measurements over-estimated their height and they were not the tallest living plants in the world. Thus, a recent article in Wikipedia entitled "List of Superlative Trees" gave the title of the tallest living tree to a Coast Redwood, known as Hyperion, (Sequoia sempervirens) in the Redwood National Park in California at 380.3 ft. or 115.92 m. In second place, however, came a Mountain Ash, known as Centurion, (Eucalyptus regnans) in the Arve Valley, Tasmania at 329.7 ft. or 100.5 m.

Boyle's claim of Eucalypts within 25 miles of Melbourne measuring in excess of 420 ft. in height therefore turned out to be inaccurate – we believe that it not through any dishonesty on his part but because of inadequacies of his method of measurement. However, Boyle's work as a plant collector for Baron von Mueller was recognised by the Australian National Herbarium which notes that "David Boyle was the collector of the lectotype (or *definitive type example of a species) of Eucalyptus regnans F. Muell. (1888)."*

Throughout his long life and particularly between 1860 and 1890, Boyle was interested in all things botanical. He was an active member, office bearer and/ or judge in the Victorian Horticultural Improvement Society and the South Bourke Horticultural and Agricultural Mutual Improvement Society. He also developed a large garden and nursery on his Forest Hill property and probably had a major influence in the development of the plant and flower industry in the Nunawading District.

Unfortunately, David Boyle died before Walter Jones purchased his Tunstall property.

APPENDIX THREE

What Are Bulbs, Corms, Tubers, and Rhizomes?

Flower "bulbs" come in these forms: true bulbs, corm, tuber, tuberous root, and rhizome. So, what you might think of as a flower bulb may not be a bulb at all — botanically speaking, that is. The following material has been copied from https://www.dummies.com/home-garden/gardening/flower-gardening/gardening-what-are-bulbs-corms-tubers-and-rhizomes/

TRUE BULBS

True bulbs, such as daffodils, tulips, hyacinths, and snowdrops, often have a papery skin or tunic on the outside, much like an onion. Bulbs with a papery covering are called tunicate bulbs. The tunic helps protect the bulb from drying out when it's resting or waiting to be planted. However, some true bulbs, such as lilies, don't have a tunic. These bulbs dry out faster and are more easily bruised.

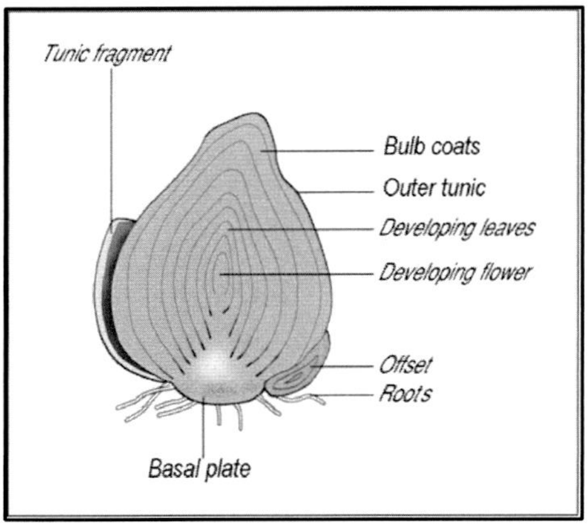

All true bulbs share the following characteristics:

- They're more or less rounded, sort of ball-like, and narrow to a point on the top. Leaves and flower stems appear from this point.
- With or without a tunic, true bulbs have a flat part, called a basal plate, at the bottom. That's where roots grow and also where shoots and scales are attached.
- True bulbs have new bulbs, called offsets, which form from the basal plate. When they get big enough, these offsets, or daughter bulbs, produce flowers on their own.
- True bulbs are made up of rings, called scales, which are modified leaves that store food. Cut apart a true bulb, such as a hyacinth, at the right time of year, and you can find a miniature flower inside, just waiting to begin growing. Perennial true bulbs

add new rings each year, from the inside. Old rings on the outside are used up, but the true bulb itself persists from year to year.

CORMS

If any of the characteristics that identify true bulbs are missing, the plant isn't a true bulb. Instead, it's a corm, tuber, tuberous root, or rhizome. Popular corms include crocosmia, gladiolus, freesia, and crocus.

Corms have these traits:

- Corms have a tunic. The tunic may be fibrous, what botanists call netted or reticulate, or the tunic may be smoother, with distinct rings, what botanists call annulate.
- Corms have a basal plate at the bottom and one or more growing points at the top. Bulbs and corms both have a definite vertical orientation.
- Corms are undifferentiated, uniform, and contain no rings when cut apart. Corms are stem tissue, modified and developed to store food.

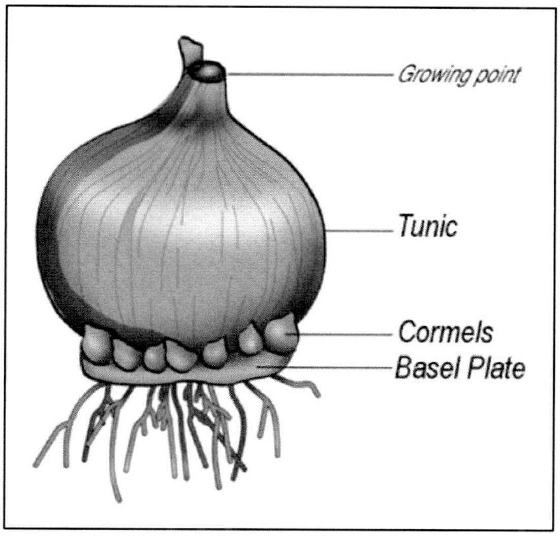

- The corm you plant is used up for growing the flower. Before it withers away at the end of the growing season, however, a brand-new corm (sometimes several new corms) forms and replaces the mother corm. The new corm contains the food reserve for the dormant crocus or gladiolus until it's time to grow again.

TUBERS

You may not know it, but you might already be familiar with a popular tuber: the potato. Other tubers include tuberous begonia and cyclamen. Tubers have these features:

- Tubers have no tunic.
- Tubers lack a basal plate. Most tubers root from the bottom.
- Tubers have several growing points, called eyes. More organized tubers, such as caladiums or tuberous begonias, have their eyes at the top. Some tubers, such as anemones, aren't so orderly. Distinguishing the top from the bottom of the tuber may be difficult. If you're not sure, plant it sideways and let the tuber figure out which direction to grow.
- Tubers are made of modified, undifferentiated stem or enlarged hypocotyl tissue. They have no highly specific internal structure.
- Tubers don't make offsets or produce new tubers. Tubers usually just get bigger each year, making more growing points.

TUBEROUS ROOTS

Tuberous roots are modified, enlarged, specialized roots that store food, and are used up during the growing season to be replaced by new storage units. The tuberous roots cluster together, joined to the bottom of a stem. The stem contains the new growing point for the next year — a piece of root alone won't grow. Examples of tuberous roots are dahlias, daylilies, and sweet potatoes.

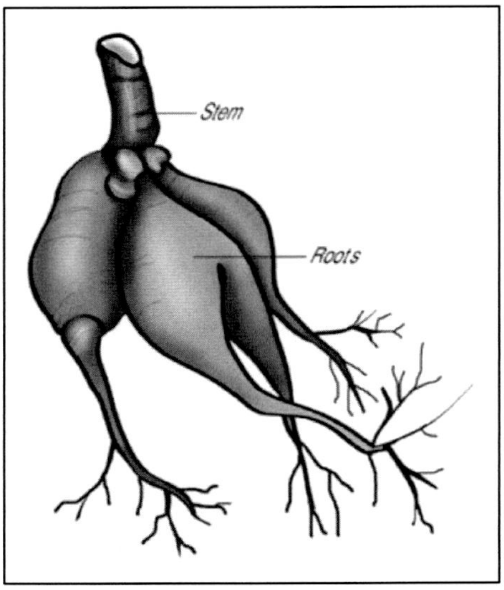

RHIZOMES

Rhizomes are stems that grow sideways rather than up, running along the surface of the soil or just below it. Plants that use rhizomes for food storage have fatter, more bulblike rhizomes, covered with a dry base of leaves. Rhizomes branch out, and each new portion develops roots and a shoot of its own. Familiar rhizomes include iris, lily-of-the-valley, canna, and ginger (Zingiber officinale).

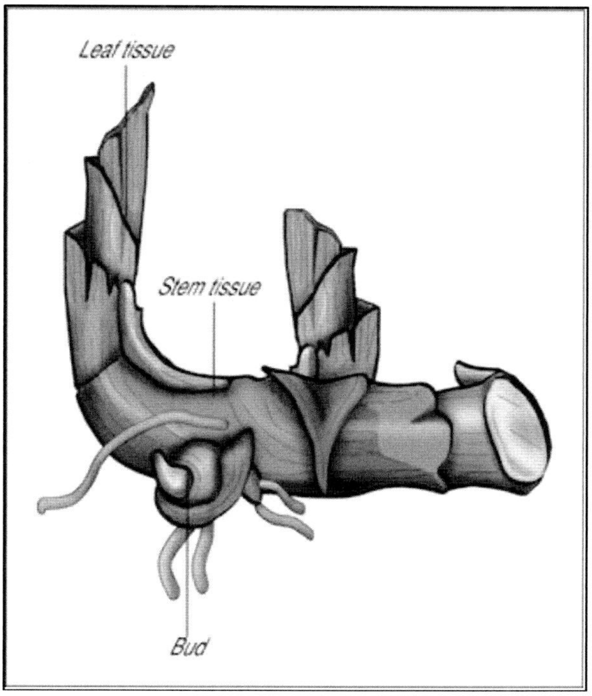

INDEX

Index page numbers in bold refer to photographic images.

Flowers or fruits from a selection of plants grown on "Rosemont"

Made in the USA
Las Vegas, NV
15 April 2026

45227085R00067